I0393867

MAY 04

Science and the Law:
2001 and 2002 National Conferences

2001 and 2002 Conference Cosponsors

National Institute of Justice, U.S. Department of Justice
American Bar Association, Criminal Justice Section
American Academy of Forensic Sciences
National Center for State Courts

In Collaboration With

The National Academies
The American Association for the Advancement of Science

NCJ 202955

Sarah V. Hart
Director

Findings and conclusions of the research reported here are those of the conference presenters or attendees and do not reflect the official position or policies of the U.S. Department of Justice or any of its offices, cosponsors, or collaborators. This summary of discussion is not a verbatim transcript; questions concerning details of the research or presentations should be directed to the speakers.

The National Institute of Justice is a component of the Office of Justice Programs, which also includes the Bureau of Justice Assistance, the Bureau of Justice Statistics, the Office of Juvenile Justice and Delinquency Prevention, and the Office for Victims of Crime.

Executive Summary

The annual National Conference on Science and the Law brings scientists, attorneys, and academicians together to examine issues that arise at the intersection of the scientific and legal professions. The two professions take distinctly different approaches and do not speak each other's "language." The need for them to communicate better becomes more imperative with further advances in science and technology.

Third Annual Conference on Science and the Law, October 4–6, 2001

The third conference, held in Miami, Florida, took place shortly after the September 11 terrorist attacks on the United States. An overall theme was how scientists can convey complex information to juries. Other themes included the development of standards for collecting and managing digital evidence, the use of law enforcement personnel as expert witnesses, and the roles of court-appointed expert witnesses.

Keynote Speaker Addresses Use of Forensic DNA

Christopher Asplen, Executive Director of the National Commission on the Future of DNA Evidence, spoke about the intersection of science and the law as it affects the use of forensic DNA. Advances similar to those in DNA evidence are likely to take place in other criminal justice-related technologies. An example is genetic profiling, which could be used to track down terrorists. Expanded use of technology is possible *because* there are experts who know when such uses are and are not appropriate.

Reports on Science and the Law

Some of the major advances in the sciences, including medicine and genetics, were cited. Legal developments in the past year included the application of the *Daubert* ruling to contexts different from the original, medical one; the large number of challenges to forensic science expert testimony; and unresolved issues in forensic psychology and psychiatry.

Reports of Conference Sponsors

Each conference sponsor explained steps it is taking to bridge the science-law gap. As the use of expert witnesses increases, the courts have been slow to comprehend changes in technology. Aids are becoming available to help judges and attorneys understand scientific matters. An attempt is being made to bring new standards of objective examination and validity to the forensic sciences. The American Association for the Advancement of Science noted it has a pilot program to help judges appoint qualified scientists as experts. The National Academies bring the science and security fields together to address such issues as foreign students and scholars.

Tutorials for Scientists and Legal Professionals

Tutorials for scientists and legal professionals offer each an understanding of the other's field. For attorneys and others in the legal profession, information was presented on the basics of the scientific method, including such concepts as the robustness of data and statistical inference; an overview of the history of forensic science; and the use of computerized databases. Tutorials for scientists included an overview of the current state of expert

witness testimony (covering the *Frye* and *Daubert* standards) and presented information on how expert witnesses are different from "fact witnesses," the preparation and trial use of scientific evidence, the role of the forensic scientist in the legal setting, and the judge's role when science-based evidence is submitted.

Conference Sessions

Forensic Fraud

Falsified evidence, which may result in culpable misconduct, is different from flawed evidence. Scientists are not to blame for a wrong conviction based on flawed evidence, but the public expects them to make no mistakes. Convictions overturned by DNA and other forensic evidence indicate that law enforcement must pay close attention to evidence collection to avoid flawed evidence. With respect to falsified evidence, laboratory protocols can insulate people from bias and the incentives for it.

Use of Court-Appointed Experts and Advisors

Federal Rule of Evidence 706 authorizes judges to appoint scientific experts to either testify or help the court decide what evidence is admissible and who should testify. But opposite conclusions can be reached with the same set of facts, and some issues are unresolved; for example, States differ over whether indigents must be provided with experts. A registry of scientific legal advisors, now in the experimental stage at a major university, contains lists of experts available to the courts. More than half of Federal judges who had used a court-appointed expert did so only once, and most of those who had not used one said they would do so only if the adversarial process was failing and they needed more information. Although a judge may want to consult an expert,

the parties to the case may not want the judge to do so. One study revealed that in almost all cases in which an expert was used, the court followed the expert's advice.

Law Enforcement Officers as Experts in Court

Police are often called on to testify on such matters as drug recognition, the criminal's modus operandi, profiling of drug couriers, and police practices. Because their expertise is based on their experience in their official duties, experience as well as training and other qualifications should be considered when selecting police as experts. Recent trends indicate juries do not necessarily view these experts favorably. Police credibility has been challenged by recent instances of alleged misconduct.

Eyewitness Evidence: Can Memory Be Improved?

Eyewitness evidence is often the most influential factor in criminal proceedings, but it can produce wrongful convictions. The police receive little training in collecting eyewitness evidence. Memory can be distorted in the criminal justice process by the use of leading questions, for example. Eyewitnesses tend to want to make a judgment even if they are not certain rather than make no judgment at all, but blind testing can help correct this. New techniques for improving memory, based on cognitive psychology, can improve eyewitness identification. The "cognitive interview" recognizes the interaction of interviewer and witness and tries to recreate the original context by asking questions that let the witness take the lead. This method has been found to elicit more information but only slightly more accurate information. More research on memory is needed to improve juror decisionmaking.

Presentation of Papers Selected From the Open Call

The following papers were presented:

- "Interdisciplinary Collaboration: The Key to Improving Medical Documentation of Domestic Violence for Use as Legal Evidence," by V. Pualani Enos.

- "'Task at Hand' Principle of *Kumho Tire* v. *Carmichael*, 526 U.S. 137 (1999)," by Michael Risinger.

- "Recent Defense Challenges to Forensic DNA Evidence," by William C. Thompson.

Mock Trial on Use of Digital Evidence

Electronic (digital) evidence is an emerging area of forensics. To convey the complexity of cases involving this type of evidence, a mock trial was held. The mock trial was based on a real murder case in which the victim was found beside a computer. The key issue was whether the prosecution's expert testimony was reliable and thus admissible. Contamination issues arose because the first responder examined the files before the computer recovery expert arrived. The defense contended that the first responder unwittingly altered the files and that obtaining information from the suspect's computer violated privacy laws. The reliability of CaseView, a program used by law enforcement to copy files, was called into question by the defense. Despite the defense objections, the judge decided that the prosecution's evidence was admissible.

Keynote Address on Jurors' Comprehension of Scientific Evidence

Professor Lawrence Solan from Brooklyn Law School described a study that examined jurors' understanding of scientific evidence. The results indicated that jurors give more weight to evidence presented in lay terms, but they also give more weight if the language is incomprehensible but the presenter has impressive credentials. Jurors also tend to react unfavorably to paid experts.

Fourth Annual Conference on Science and the Law, October 3–5, 2002

The theme of the 2002 conference, held in Miami, Florida, was emerging trends in scientific evidence, including new issues in forensic DNA, ethical issues facing forensic scientists, how attorneys can interpret scientific reports, genetic research, the role of scientists in counterterrorism, digital evidence, and the reliability of fingerprint evidence.

Preconference Workshops

Interpretation of Scientific Analytical Reports (for Lawyers)

The workshop was intended to help attorneys interpret scientific analyses in reports. The analyses often follow certain standard models to guide data gathering and interpretation. It is essential that the report specify the analytical method used. Both the writer and the reader of the report are responsible for interpreting it correctly. However, conclusions, particularly "inconclusive" results, can be interpreted differently. Attorneys must ascertain that the laboratories issuing the reports are accredited. Laboratories that prepare DNA and serology reports should have "toolkits" that include sets of guidelines and procedures manuals. A hierarchy of terminology on probability is being developed for use in the preparation of document examination reports. Mental health assessment reports must identify forensic issues and legal questions; they are not intended for therapy.

Brady and Other Ethical Issues Facing Forensic Scientists

Much evidence acquired by prosecutors may be material to the defense. The 1963 *Brady* v. *Maryland* decision requires them to turn over potentially exculpatory information to the defense. *Brady* is sometimes seen as asking the prosecutor to aid the accused. It has produced more Freedom of Information Act discoveries by defense and more attempts to find out about misleading evidence. One presenter noted that defense counsel needs adequate breadth of discovery to obtain scientific evidence. On the other hand, *Brady* has in some cases led to large additional areas of discovery for information that is only circumstantial.

Can DNA Be the Magic Bullet? What DNA Can (and Cannot) Do

Issues in the use of DNA evidence continue to emerge. Among them are whether there is a right to postconviction relief based on DNA, the scientific limitations of DNA testing, and the inability of many crime laboratories to work every case that involves DNA evidence. Analytical problems persist even though information expands. Computer-assisted data interpretation can help reduce laboratory backlogs. One presenter noted that the common assumption that DNA evidence wins the case could be dangerous. Defense attorneys sometimes do not ask for independent DNA testing because problems like contamination can arise. Although the *Daubert* decision required assessing evidence for its admissibility, courts still have not decided how to treat mixed-DNA evidence.

Keynote Address on DNA and Genetics: A Challenge for Lawyers and Judges in the New Millennium

In science, there is a distinction between "error" and "mistake"; in the law, there is no such distinction. When a mistake occurs in a scientific experiment, the experiment can be conducted again. Errors in experiments need only be documented. In the law, an error is the same as a mistake because it may overturn a decision. Exoneration via DNA has become fairly frequent, but DNA databases remain controversial. As genetics research continues to shed light on these issues, it is likely to have more influence on the law. The discovery of genetically caused diseases may raise issues of privacy and classification of people by their DNA. Medical information is already being used to make some hiring, firing, and promotion decisions.

Reports on Science and the Law

Daubert is not the only evidentiary standard, and the sky may not be falling as a result of it. Peer review is a standard, although one on which not too much emphasis should be placed in the legal context. Changes in technical fields affect testimony, including police officers' testimony and clinical medical testimony. The *Kumho Tire* decision illuminated the issue of rigor in a variety of technical fields, causing, for example, handwriting evidence and fingerprints to be increasingly challenged. Typically, police are not asked to explain the basis of their experience when they testify, but scientific experts are asked to do so. Certain issues have created essentially a scientific revolution in the courts. The current confusion over litigation-sponsored science is likely to promote more research that will resolve issues now in conflict.

Tutorials for Scientists and Legal Professionals

Three tutorials were presented: basic rules for using expert evidence, preparing expert testimony, and toxic torts.

Expert Evidence: Basic Rules and Contemporary Controversies

Daubert and *Kumho Tire* require judges to ascertain that scientific evidence is relevant and reliable. How do they affect the analysis of fingerprint, handwriting, and trace evidence? *Daubert* criteria have been used almost as a definitive checklist. *Daubert* and *Frye* have asked the courts to be less deferential to scientists on the basis of their credentials alone. As a result of these two cases, pretrial hearings and hurdles for admissibility have sometimes increased. But many forensic sciences have not come out of a university-based research tradition of controlled experiments.

Investigating, Evaluating, and Preparing Experts

The credentials of the expert need to be verified in view of documented misconduct by expert witnesses. This documentation has led to increased scrutiny and accusations of negligence, fraud, or even criminal malfeasance. Judges and jurors are becoming more mistrusting of expert witnesses and peer review standards, and many more hearings are being held on expert witness evidence. For these reasons, it is incumbent on attorneys to prepare experts for trial.

Toxic Torts

Scientists often feel ill at ease in legal proceedings because they are not familiar with adversarial situations and feel reluctant to support only one side of an issue. The law, however, avoids ambiguity. In torts involving chemicals, scientists may legitimately differ on how they rate the health effects of a toxic substance. Risk assessment is increasingly used in court cases, as in government regulations.

Plenary Panels

Plenaries were presented on science and counterterrorism, the manipulation of digital evidence, and jurors' weak grasp of statistics.

Counterterrorism

Science has a role in securing the homeland against terrorism. A proposal now before Congress would create a national research and development enterprise focused on homeland security that would involve many fields, including chemistry, physics, and the life sciences. Some scientific developments will be used in security measures; biometrics will affect travel document verification, for example. Another example is the development by the Federal Government of new diagnostics to detect pathogens affecting humans, animals, and plants and new vaccines to prevent infections from these pathogens.

When Is Evidence Considered Manipulated? A Close Look at Digital Evidence

Nearly every kind of case now has some connection to digital evidence. Computers can be crime victims, crime instruments, or crime evidence. In the absence of eyewitnesses, chatroom and e-mail files, for example, can be used to establish a connection to a crime. Until recently, however, little training has been available in investigating computer crime. Skillful image manipulation is very hard to detect; it is often difficult to distinguish manipulation from normal anomalies in files. Defense often seeks exclusion of computer evidence as hearsay, but courts will admit exceptions if the evidence is generated as a result of the ordinary course of business.

Juries' Understanding of Statistics

Jurors are neither overawed by nor dismissive of expert testimony, according to a study of the videotaped deliberations of juries in civil cases. Jurors frequently discussed the experts and focused on the content, plausibility, and "expertness" of their testimony. They disliked obfuscation. Jurors are unlikely to understand statistics, at least not as well as are undergraduates who take statistics courses and have backgrounds in mathematics. Study design is an important concept to convey to juries, as it determines the reliability of the conclusions. But one study indicated that only about half the jurors understood problems with a nonrandomized study.

Presentation of Papers Selected From the Open Call

The following papers were presented:

- "The Role of Meta-Analysis in the Legal System," by Jeremy A. Blumenthal.

- "Voodoo Science by Default," by Peter R. DeForest.

- "Inconsistency in Eyewitness Testimony: What Does It Really Tell Us?," by Ronald P. Fisher.

- "Jurors' Comprehension of Contested DNA Evidence: A Case Study," by William C. Thompson.

Fingerprints: Making Sense of Forensic Science—Plenary and Roundtable Discussion

Current court treatment of fingerprint evidence may be a keystone for treatment of other kinds of forensic evidence. Despite challenges to fingerprint evidence, much of it is admitted on the basis of experts' statements. Courts should be better able to filter out statements that are not based on sound research. Lack of research could threaten forensic scientists' credibility. The main issue is reliability, but science is not reliable; rather, it has rules for the reliability of *testing*. Thus, fingerprint evidence might be admitted as technical expertise, not as science. *Daubert* recognized the limitations of science. A requirement for scientific evidence at all times would handicap the courts. However, it is difficult to know what nonscientific, technical standards should be used for trace evidence.

Keynote Address on Knowledge, Power, and the Evolving Role of Scientific Evidence

The Honorable Gerald T. Wetherington elaborated on the mental struggles of juries, judges, and expert witnesses as they try to make the right decision. In the murder case he cited as an example, an overheard telephone conversation was presented as evidence. Overhearing conversations can be illegal, which raises the question of whether the evidence is admissible. Another issue is subjective factors that can influence a decision. Expert witnesses need to confront their own observer bias and scrutinize their motives for testifying.

How Much Can the Hair Tell? Microscopic Examination

In a mock legal scenario, a defense motion not to accept hair as evidence was denied. The defense challenged on several bases. An expert witness provided an overview of the value of hair as evidence, explaining how unknown and known hairs are compared microscopically to try to match them. However, hair evidence is not easily quantifiable or digitized and reliability and validity are not as readily achieved as for DNA evidence. Another expert spoke in favor of mitochondrial DNA (mt-DNA) testing of hair. If microscopic comparison produced a match with the defendant's hair, mt-DNA testing might produce a different result.

Contents

Third Annual Conference on Science and the Law, October 4–6, 2001

Thursday, October 4

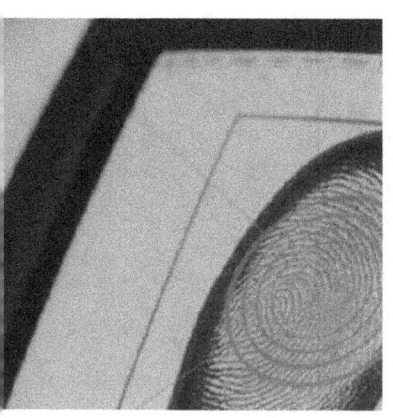

Welcome and Opening Remarks

Sarah V. Hart, Director, National Institute of Justice (NIJ), Office of Justice Programs (OJP), U.S. Department of Justice (DOJ), opened the conference by reviewing NIJ's work for the participants: researching policy issues for public safety and law enforcement, staging demonstration projects on promising approaches against crime, and developing science and technology for justice programs. Following the disastrous events of September 11, 2001, NIJ's Office of Science and Technology (OST) arranged to have scientists in the field in New York City to help with rescue efforts using the newest available technology, such as miniature cameras strapped to rescue dogs searching in the rubble. There was a great sense of mission as DOJ staff, Federal Emergency Management Agency (FEMA) personnel, and State and local rescue workers struggled to save the people they could reach.

Ms. Hart said debate is inevitable when the disciplines of science and the law are placed together. Both the rule of law and the protective technology that supports it are necessary for society. This timely conference generates ideas for common goals to ensure both public safety and essential liberties. She quoted Benjamin Franklin, signer of the Declaration of Independence, who said, "They that can give up essential liberty to obtain a little temporary safety deserve neither liberty nor safety."[1] She said we must use

technology to its greatest potential without interfering with fundamental fair process. Science and the law have somewhat conflicting approaches. Science, as a profession, values change and innovation— "thinking outside the box." The legal profession often takes an opposite approach, encouraging stability and reverence for rule of law. The law typically is not designed to encourage innovation by courts or judges.

New requirements in the changing world challenge both of these disciplines. An example is the procedure of requesting wiretap orders from a judge. Previously, this required law enforcement officials to seek a warrant using both a phone number and a name. In the case of terrorists, however, cell phones or disposable phones often are used, and there is no ability to respond quickly. Under new legislative proposals that warrants can now be obtained using only a name, the law enforcement officials still must make a case to a judge for an impartial determination. Newer technologies continue to raise these debates about how the law should change to accommodate legitimate law enforcement needs while still protecting constitutional rights.

The Supreme Court, during its last term, found against the prosecution's use of a technology that sees through walls to find "hot spots" (e.g., possibly marijuana plants). The Court would not allow law enforcement to perform this kind of search without a warrant, concluding that it invaded basic privacies. Members of the

Court were very divided over the issue. Ms. Hart said that scientists and lawyers particularly need to resolve communication issues and understand each other better.

David G. Boyd, Deputy Director, NIJ, and Director of NIJ's Office of Science and Technology, said the previous National Conference on Science and the Law focused on admissibility of evidence when complex scientific technology was involved. This year's conference adds the aspect of talking to juries. Lawyers and scientists both use special languages, and juries speak neither language. In addition, lawyers and scientists may be interacting only in an adversarial situation. How can this rift between science and law in such a setting be bridged to increase necessary understanding? The conference sessions examine cutting-edge issues in science and law, such as the development of rational standards for the collection and management of digital evidence, the use of experienced law enforcement personnel as experts, and the roles of court-appointed expert witnesses.

Keynote Address: So, Tell Me, Is DNA Really That Good?

Christopher H. Asplen, Executive Director, National Commission on the Future of DNA Evidence, said the Commission's work has promoted public discussion and has provided an open forum for consideration of the issues related to the increasing use of DNA evidence. The process has helped foster better policy decisions that affect the provision and use of the resources for both the development of DNA technologies and the protection of society through the justice system. Mr. Asplen pointed out that science and law have an uncomfortable, "unnatural" relationship, but the need to integrate these

different viewpoints has accelerated at a breakneck pace.

In the last 4 or 5 years, DNA technology has developed and been integrated for investigative purposes at a rate that is "beyond the wildest dreams" of even the professionals who presented at the First International Conference on Science and the Law in San Diego. Three years ago, presenters spoke about using DNA to solve rape cases and suggested that there may be errors in existing convictions. At that time, listeners' jaws dropped in shock. Now, postconviction DNA testing has legislative support in 25 States. Similarly, only 3 or 4 years ago, the FBI alone used computer databases containing DNA information to investigate homicides and sex crimes. Last spring, more than 115 pieces of legislation in States across the country expanded development of DNA databases for convicted offenders.

There is still a substantial problem with providing supporting resources to the States, where large sample backlogs exist. However, last year Congress approved $170 million for backlog reduction in the States, and outsourcing laboratory analysis of crime evidence has become a working solution for about half of the backlogged samples. As databases continue to improve, the power of this DNA technology is being used more routinely to investigate cases that were previously unsolvable. Genetic profiles are becoming more reliable, and law enforcement personnel now sometimes file "John Doe" criminal warrants based on them. Arbitrary statutes of limitations have been set aside in circumstances in which DNA analysis 10 or 15 years after the crime was able to prove innocence.

Mr. Asplen said we should expect similar acceleration in other technological advances. There will be new ways to track down terrorist crimes, and an open forum to inform broad criminal justice policies is

needed for those technologies as well. Consider, as an example, the progress in tracing allele sequences in genetic profiles, which enables the investigator to discover the geographic background of the person tested (such as ethnic/biological connections to Asia, Ireland, or the Middle East). Should we think of this, asked Mr. Asplen, as a pitfall or an advantage? What if there were a warning call about an explosion, and DNA was recovered from the phone from which the call came? Should we use the DNA results for allele sequencing? He asked listeners to think about the time, money, and anguish that would be saved in responding to and investigating large disasters as missing persons are located and identified. Mr. Asplen said we can talk about expanded police powers and use of technology (such as wiretaps, Internet search capacity, and interception of cell phone communications) particularly *because* we have people who will "put the brakes on" when appropriate. Attention to the Constitution and protection of people's rights reasonably balance such initiatives.

Annual Reports on Science and the Law

Anticipating Science Impacts— Updates on the Last Year and the Next Decade

Martin A. Apple, President, Council of Scientific Society Presidents, discussed scientific advances. The scientific associations represented by Dr. Apple's council publish millions of scientific articles. He suggested to attendees that his discussion could only touch on some of the most rapid advances. He offered to extrapolate to some provocative, possible future scenarios. People who occupy leadership roles today must be lifelong learners, he said, because the application of science (with rapidly changing areas of knowledge) to crimefighting is like "hitting

a moving target." Revolutionary innovations that would have been unbelievable in the first half of the 20th century have become accepted and even necessary (e.g., synthetic hormones, antibiotics, manned space stations, cracking the human genetic code, babies conceived in test tubes, electricity without resistance, and crossing the ocean in 2 hours).

The agricultural demand for food production to support world population has required capturing most usable real estate on the Earth, Dr. Apple said. Genetic engineering has made possible a greatly increased food supply. In a recent example, researchers at University of California–Davis altered plants to enable them to grow in even high-salt environments. Sophisticated surgical procedures like heart bypass that require specialized training and experience can now be conducted at great distances using robotic systems. Recently, a New York surgeon was able to remove a gall bladder from a patient in Salzburg, Austria, using such technology.

Other notable advances include the following:

- The discovery in Boston of a human gene that confers longevity; confirmation of a "learning enhancement gene" in mice.

- The invention of switches and circuits that use photonics (photons instead of electrons), enabling the construction of a "quantum computer" with unimaginable speeds.

- The development of fuel cells that will use hydrogen (perhaps by splitting water molecules) in a process that has water as an end product (no net change and no net pollution).

- The results of studies of the exact cell receptors for medical drugs that are making the drugs many times more potent and will result in fewer side

effects; however, studies will be needed to monitor the environmental impact of excreted biological agents from such designer or "miracle" drugs.

- The use of stem cell research to speed up the development of several anti-AIDS vaccines (now in trial). The next stage will be the growth of replacement organs.

- MIT's recent design of a modular "pebble bed" nuclear reactor that controls just enough fuel to start the chain reaction but cannot accumulate a dangerously large amount in one place.

- The development of new ultrasecurity locks that use a nanoscale maze that jams in the case of any error.

- The synthesis of high-temperature plastic-ceramic materials that can replace many conventional hard materials.

Dr. Apple pointed out that society has no consensus or guidelines concerning the moral or religious implications of many of these advances. Strange social questions could arise in the new scientific contexts (for example, Is a reproductive clone a person's sibling or son/daughter?).

Annual Report on Science and Law

David Faigman, Professor, Hastings College of Law, University of California, spoke about recent courtroom use of expert testimony and referred to a survey of 700 Federal cases involving the Federal Rules of Evidence and the *Daubert* decision.[2] Although the substantive area (expert testimony) is complicated in itself, some major issues also transcend the subject. Courts are starting to use *Daubert* in an administrative context, i.e., as a style of evaluating expert witness testimony. Federal Rule of Evidence 702, which was supposed to codify *Daubert,* actually

introduced new elements. The idea, for example, that the data must have sufficient "fit" for the inferences being drawn obliged the Court to screen expert testimony both for sufficient factual basis and applicability to the facts of a given case.

In a similar situation, but in a clinical medical context, the idea of "differential diagnosis" has been important. In a medical case, such as an allergy, doctors are interested in the cause for research (not just in treating the symptom), so differential diagnosis is accepted. Courts have been reluctant to generalize from the specific symptom or disease event to the general etiology or root cause, but sometimes this is done implicitly. Differential diagnosis has also become important for studying cause in an engineering context. The revolution brought about by *Daubert* relates to introducing the same sophistication to other forensic sciences. Base rates, probabilities, regression analysis, and error margins are now brought up in consideration of such different questions as "battered woman syndrome" or traces of bite marks.

In a court case, however, it must be clear what question is being asked: Some courts give value to temporal proximity, but how far should that be relied on in the question of the cause of a crime? How much epidemiology is really necessary to give "sufficient data"? Many women with tissue and autoimmune disorders pursued the class action case involving silicone implants. Is it enough to go to a jury with just clinical anecdotal information? Engineers have faced the greatest difficulty in recent cases because they have often been met with expectations that they should be able to rule out alternative explanations, explain methodology, know effects of "confounding variables," and so forth. Courts must be innovative enough to examine different areas in different expert contexts.

Professor Faigman said there is currently a surfeit of challenges to forensic science expert testimony. He quoted Oliver Wendell Holmes, who said, "No facts make bad law." Some evaluation of forensic technical work has been badly done in such areas as fingerprinting. There is also still a wide range of unresolved issues in psychology and psychiatry. The Supreme Court often delegates matters dealing with science, child witnesses, and polygraphs to specialists. "For my part," said Professor Faigman (considering *Daubert*), "if a doctor used a stethoscope to diagnose cancer, you could say it was a valid method but that method would not apply properly to the case in hand." Judges have a responsibility to examine both methods and conclusions, not just to "trust experts." The court has to ask, "How did you examine it? Explain how the premises compel or support your conclusion." He noted that for centuries, medical expertise supported the use of leeches, but scientific experience grew beyond that.

Professor Faigman said he considered the challenge to convince mainstream scientists to research mainstream legal issues as most important. Primarily, there has been neither enough prestige nor financial support. Why, for example, are researchers not looking at polygraph testing? In the case of police expert testimony, the courts have not been too good at determining the basis for conclusions. They need to do somewhat better than simply asking, "In your experience, do drug dealers carry guns?" The district courts now have to supply reasons for either admitting or excluding expert testimony. Standard appellate review may remedy the lack of appropriate expert testimony or may raise questions that transcend the local dispute. There is now a tremendous pressure on the appellate courts to settle scientific evidence disputes.

Open Discussion

One participant asked why courts often ignore methods that are "clear rules" in scientific procedure. In the *Van Wyk* case, for example, the court admitted as evidence threatening letters said to be from a defendant, even though there had been no discussion of validated linguistic techniques, error rates, peer review, and so forth.[3] Professor Faigman noted that we are in a revolution or transition period, and many examples exist on both sides. Some courts are "without a clue." Hopefully, with time, this will change. He agreed with other comments that law schools are not teaching enough about hypothesis testing and scientific methods.

Another participant, who said he was a judge and former scientist, noted a failure to communicate fundamentals of science that would enable lawyers to distinguish "pseudoscience" and "quasi-science" (such as a police officer discussing fingerprints). Very seldom does legal education offer instruction in statistics or information on how to extrapolate data.

Dr. Apple mentioned ongoing disputes in the scientific world, such as disagreements about the dose of ionizing radiation that would be dangerous, and said scientists do ask for government priorities. However, dictating the direction of scientific research does not give the best results either.

The group discussed limitations of the current adversarial system. Professor Faigman said the jury system is usually cited as the component that gives rise to the adversarial system. Many rules limit what the jury sees. Too many judges and lawyers consider this form to be sacred. Rather than preserving impartiality, the attitude often prevents intelligent use of science and technology. Insights could come from a "civil factfinding" or "inquisitorial" model. Jury trials are an aspect of

the system that has possibly grown out of control and may need overhaul. Professor Faigman noted that the real revolution of *Daubert,* however, is the mandate to collect the data.

Annual Reports of Sponsors

National Center for State Courts: Assisting Courts in the 21st Century

Judge B. Michael Dann, formerly a trial judge in Maricopa County, Arizona, and now Visiting Fellow for the National Center for State Courts (NCSC), spoke about assisting courts in the 21st century. Courts have been slow to adapt to changes in technology and in the nature of caseloads. In the present era, a court case that is "worth litigating" can be expected to bring two or three expert witnesses to the stand. NCSC has launched an online resource center for judges to help them consider expert witnesses, including material compiled for the Federal Judicial Center and the National Judicial College. The site is organized to be friendly to judges, lawyers, and law clerks. Articles are included on scientific methods, culture, statistics, and surveys, and links to an online encyclopedia and annotated treatises on science and law (Westlaw) are provided.

The center is also discussing a project with Duke University, which would use the university's Registry of Independent Scientific and Technical Advisors (part of Duke's Private Adjudication Center [PAC]). Judge Dann said that a sitting judge does not have time to surf the Internet for research sources, and States often cannot pay to send their judges to special training. The new project's concept would develop on-call "mentors" in particular fields to tutor judges in background science and provide quick answers for certain technical questions.

Judges have been reluctant to interfere in party control of litigation or to stimulate a "satellite litigation" concerning experts. Guidelines could clarify which tasks are to be performed by experts, what to tell the jury about the status of the technical master, and what to do with the report. Experts may also tutor the judge in a scientific area to provide the judge with background information before the judge hears an expert witness. Typically, State trial judges do not even have law clerks to help with research. Much remains to be studied on scientific and technological issues that are affecting the courts.

American Academy of Forensic Sciences: Building the Bridge Between Forensic Science and the Law

Mary Fran Ernst, President, American Academy of Forensic Sciences (AAFS), and Director, Medicolegal Education, St. Louis University Medical School, said her umbrella organization represents more than 5,000 professional forensic scientists in more than 18 specialties. One of their important activities has been establishing independent forensic science accreditation boards to bring new standards of objective examination and validity to these disciplines.

The association also works a great deal in the field of science education. Ms. Ernst referred to recent studies that show declining scores in mathematics and science for American children of middle school and (particularly) high school age. Her organization has a grant to bring forensic science into high school classrooms to engage the students. AAFS participates in a DOJ Technical Working Group for educational standards of people entering the forensic field.

AAFS produces a world-class journal, Ms. Ernst said, that offers bimonthly information on new work in the field of forensic

science. The association's annual meeting is attended by more than 2,500 scientists working in forensics. Ethical issues, Internet use, and computer forensics were among the interesting panel topics at the last membership meeting in Seattle. The upcoming year's theme will be accreditation, education, and professional integrity.

American Association for the Advancement of Science: Law and Science Initiatives

Mark S. Frankel, Director, Program on Scientific Freedom, Responsibility and Law, American Association for the Advancement of Science (AAAS), described his organization as the largest multidisciplinary scientific association in the world. It publishes *Science,* a peer-reviewed weekly journal. Current initiatives include the following:

- Improving science education at the K–12 level in schools.

- Helping scientists in the former Soviet Union develop a competitive research funding system.

- Establishing ethical guidelines for further developments in science.

AAAS recommends witnesses to Congress, issues statements to the public, and works in many collaborative efforts. Since 1974, it has participated in a joint committee with the American Bar Association (ABA) to examine the intersection of science and the law. Some activities of the joint committee include providing recommendations for a legal framework for the Internet, establishing regulations that affect research misconduct, and assessing advances in such areas as genetics research. AAAS also hosts the Court-Appointed Scientific Experts (CASE) project, a 5-year pilot project to help Federal judges appoint qualified scientists, engineers, and health professionals as court-appointed experts. In the pilot approach, the judge is provided two or three names to use for a case. The expert fills out conflict-of-interest forms and is advised about what to expect in the legal arena. The CASE project has been endorsed by Supreme Court Justice Breyer and is funded by two private foundations.

Additional important work of the association includes an upcoming summit conference on future directions of science and criminal law. This meeting will bring together scientists, attorneys, and key State and Federal policymakers to develop research agendas for the forensic sciences. Many areas could benefit from more rigorous research, evaluation, and peer review as such issues as admissibility of evidence are confronted.

The National Academies: Science, Technology, and Law Program

Anne-Marie Mazza, Director, Science, Technology, and Law Program, The National Academies, Washington, D.C., told attendees that the National Academies incorporate the National Academy of Sciences (NAS), the Institute of Medicine, and the National Research Council. This nonprofit research group was founded in 1863 by Abraham Lincoln and carries out its work through studies and workshops, using members and thousands of volunteers. The Panel on Science and Law was formally established in 1999 and particularly examines science in the courtroom and its effects, and how law and judicial use of science and technology in turn affect research. The panel also addresses ethics for experts, access to research data, and Federal funding support for research.

NAS officials convene small meetings of leaders in security, science, and technology to address concerns about international scientific communication and national

security. A variety of questions have been asked about foreign students and scholars. NAS encourages a free flow of ideas and information. Draft legislation is circulating about visas and scholarships for scientific students. Further information can be viewed at *http://www.nas.edu/stl* in the coming months.

The American Bar Association: Criminal Justice Section

Thomas C. Smith, Director of the American Bar Association's Criminal Justice Section, spoke about the structure and membership of the section. He said that judges and lawyers (both prosecution and defense) are represented among its membership, in addition to court administrators, law professors, law enforcement personnel, and justice administrators. He emphasized that associate memberships are available for persons who are not lawyers but who work in the criminal justice field.

He gave some examples of the ways nonlawyer associate members can be active in the ABA Criminal Justice Section. He noted that the section regularly publishes articles from nonlawyer authors in its award-winning magazine, *Criminal Justice,* and that it has a book publishing program that may be of interest to nonlawyers who want to publish a book on a contemporary criminal justice topic. Generous royalty arrangements are made with authors who publish under this program.

He said that the section is eager to form collaborative relationships with persons of scientific ability. He pointed out that, in the past, the section has been the sponsor of study projects carried out by researchers. In such instances, he said the researchers conduct their studies independently and that the section takes care not to influence the integrity of their work and the conclusions of the studies. One recent valuable study examined "no-drop" policies in prosecutors' offices that handle domestic

violence cases. The study found that with the no-drop policy, which required that the case be prosecuted once a complaint was filed, regardless of whether the complainant continued to want the complaint pursued, alleged victims lodged significantly fewer complaints. Incident reporting also declined; partners chose to be silent rather than to allege abuse because abused partners knew that if charges were filed they could not later withdraw the complaint on which the charges were based and the charges would be prosecuted.

Mr. Smith also gave some current examples of projects being undertaken by the section that might be of interest to persons in the scientific community. These include *Electronic Surveillance Standards* and *Technologically-Assisted Physical Surveillance Standards, Guidelines for the Fair Treatment of Child Witnesses in Cases Where Child Abuse Is Alleged,* and *The Child Witness in Criminal Cases.* He also referenced the May 2002 National Conference on Cybercrime, expected to include panelists from scientific disciplines and the legal community.

The National Institute of Justice: Structure and Support

Anjali R. Swienton, Senior Forensic Analyst, ACS Defense, Inc., Contractor, Investigative and Forensic Sciences Division, Office of Science and Technology, NIJ, spoke about NIJ's structure and the Office of Science and Technology. NIJ is one of five Bureaus of the Office of Justice Programs and operates as the research organization of the U.S. Department of Justice. Although there are no laboratories at NIJ, it interacts with laboratories around the country. OST administers Technical Working Groups (TWGs) on important current areas of scientific research. There has recently been a TWG publication on digital evidence and an interactive video prepared for emergency first responders. Other topics studied

include teleforensics, databases for pattern recognition, and entomology for post mortem interval identification.

In response to the terrorist attack in New York, NIJ deployed staff to supervise integration of technology, contacted grantees who had technologies that might be useful, and arranged technical assistance to the site within 24 hours. NIJ, together with the Armed Forces DNA Identification Laboratory, is still giving systems support to New York State to analyze samples for identification of victims. This process is expected to take years. For some of the new technologies that are still being tested, the event represented "trial by fire." Court challenges might be expected in connection with aspects of their use after the disaster.

Tutorial: "Law 101"

Law of Expert Witnesses

Professor Michael J. Saks, College of Law, Arizona State University, provided an overview on the law related to expert witnesses. He discussed the differences between fact witnesses and experts. Fact witnesses are qualified by their personal knowledge. In essence, they provide a description. In contrast to the fact witness, an expert witness can give an opinion or draw inferences. The expert also must pass a higher threshold to testify as to his or her opinion. The expert witness could be helpful to the factfinder but could also be misleading. The fact witness does not have the same level of expertise as an expert witness and thus cannot evaluate expert information independently.

Professor Saks also provided a brief history of the use of expert testimony in the courtroom. In the pre-*Frye* era, the courts used the marketplace test.[4] In other words, they considered whether *consumers* of the expertise believed that the experts

were able to do what they claimed. Under the *Frye* standard, the test was whether *producers* of the expertise believed that they were able to do what they claimed. Under *Daubert,* the test has shifted to whether or not judges find that the experts can do what they claim. *Daubert* also provides a new range of criteria for expert screening, including rate of error, standards, and general methodology testing. Professor Saks provided a handout on questions that expert witnesses should be prepared to answer under a *Daubert* hearing. Examples include the following: What are the essential theories on which this field stands? What studies have been done? What is the evidence that these foundations are valid? What reliability and validity of data are associated with the expert's diagnosis and analysis in this case?

Professor Saks also indicated that the courts have not applied rigorous screening tests to such traditional areas of technical expertise as analysis of fingerprints, handwriting, and bite marks.

Preparation and Trial Use of Scientific Evidence

George "Woody" Clarke, Deputy District Attorney, Office of the District Attorney, San Diego, California, talked about the preparation and trial use of scientific evidence. He advocated a pretrial conference between the attorneys and the forensic scientist. The attorney needs to understand the scientist's methods, procedures, sampling techniques, and results. The attorney must help the scientist prepare for direct testimony and cross examination. The attorney also should work together with the scientist on using presentation technology for courtroom testimony (e.g., PowerPoint slides). The adage that one picture tells a thousand words is very true before a jury in a courtroom setting.

Epistemology of Science and the Application of Scientific Principles

José R. Almirall, Director, Forensic Science Graduate Program, Assistant Professor, Department of Chemistry, Florida International University, discussed the background of the application of scientific principles in a court of law. He described how the basic sciences relate to forensic science. He noted that forensic science takes theories of the sciences and applies them to the law. He briefly reviewed the history of forensic science and noted that the field has become much more standardized in its methodologies and interpretation of results. This evolution has been aided by guidelines from scientific working groups.

In recent years, more forensic scientists have become certified (e.g., American Board of Criminalists), and crime laboratories have become accredited (e.g., with the American Society of Crime Laboratory Directors). Dr. Almirall also noted that the use of DNA technology continues to improve and to lead to advancement in the forensics field. He noted that there must be significant improvements in the collection, transportation, and storage of evidence for DNA testing.

Function of the Judge and Jury and Burden of Proof

Judge Ronald S. Reinstein, Superior Court of Arizona, and **Judge B. Michael Dann,** Visiting Fellow, National Center for State Courts, delivered this presentation on the functions of the judge and jury and burdens of proof when using experts. The judge's role is to be the gatekeeper and decide what jurors can hear. The judge also must help qualify the expert and ask whether the testimony is reliable. Will the testimony assist the jury in understanding the evidence and in their deliberations? The jury's role is to decide whether the

evidence is credible and reliable and to give it weight in the present case. It is still unknown, however, whether juries give too much weight to expert evidence.

Judge Dann noted that evidentiary hearings have to be scheduled well before trial to decide whether to admit expert testimony. Many attorneys need far more education and training to be able to use forensic evidence properly and to their clients' advantage. Experts often talk "over the heads" of jurors. It may come across as if the attorney and the expert are having a rehearsed conversation just between themselves.

Tutorial: "Science 101"

Barry A.J. Fisher, Director, Scientific Services Bureau, Los Angeles County Sheriff's Department, introduced and moderated the session on science. Currently, court examiners of facts are concerned about reliability of procedures and distinguishing which issues need the help of a forensic expert. Defense counsel often want to know how certain the expert is in terms of numbers and probabilities.

Differing Cultures of Practice: Science, Law, and the Judicial Gatekeeping Role

Sophie Gatowski, Assistant Director, Research and Development, Permanency Planning for Children, National Council of Juvenile and Family Court Judges, discussed last year's findings on State court judges' understanding of scientific expert testimony, which were published in *Law and Human Behavior.*[5] A representative sample of 400 State trial court judges from various geographical areas and bench practices was given a 55-minute telephone interview followed by a written questionnaire. The survey gathered the judges' views of the value and intent of the *Daubert* case ruling on scientific

evidence and tested their understanding of the scientific concepts of falsifiability, error rates, peer review/publication of scientific theory, and general acceptance. The survey informed the development of a judges' benchbook on the basic philosophies and methods of science (see *www.unr.edu/bench*).

Of the judges surveyed, 205 were from *Daubert* States and 195 came from non-*Daubert* States. About 91 percent believed that the gatekeeping role for the trial judge was appropriate. Those judges who objected to the role noted that they lacked sufficient training in science and scientific methods to perform the gatekeeping function as articulated by *Daubert*. The majority of the judges surveyed reported that the idea of falsifiability was useful in relationship to expert scientific evidence, but about 35 percent did not understand the idea clearly. Most also said that the concepts of probability and error rates were valuable, but only about 4 percent had a clear understanding of these concepts. Peer review and general acceptance in the field were well-understood concepts among the judges.

Shirley Dobbin, Assistant Director, Research and Development, Permanency Planning for Children Department, National Council of Juvenile and Family Court Judges, provided an overview of some of the fundamental concepts and processes of the scientific method and hypothesis testing. To ensure that judges and attorneys are critical consumers of the expert evidence proffered in courtrooms, Dr. Dobbin underscored the importance of ensuring that they have a working knowledge of scientific methods and principles.

The scientific method seeks to organize information about the world systematically and, in so doing, discovers relationships among natural phenomena; scientific research endeavors to explain why phenomena occur and how they are related;

and scientific explanations must be formulated in a way that makes them subject to empirical testing. It is important for judges and attorneys to recognize that systematic observation and testing can be accomplished using a wide variety of methods and that the scientific method involves a wide array of approaches and is better seen as an overall perspective rather than a single, specific method. Scientific research projects typically involve either quantitative or qualitative research designs.

Quantitative research is an inquiry into an identified problem, based on testing a theory, measured with numbers, and analyzed using statistical techniques. The goal of quantitative methods is to determine whether the predictive generalizations of a theory hold true. Quantitative methods usually involve either experimental or quasi-experimental research designs. Experimental designs are characterized by random assignment of subjects to experimental conditions and the use of experimental controls. Quasi-experimental studies share almost all the features of experimental designs except that they involve nonrandomized assignment of subjects to experimental conditions and the researcher has less control over the research environment.

With quantitative methods, the experimenter frames the research question in a clear description of the target population and a logical connection to the experimental questions, testing the "null hypothesis" (no casual relationship between the studied elements). Subsample selections must be properly representative, and controls have to match the sample as closely as possible. Quantitative studies are subject to both type I errors (the null hypothesis is rejected although it is actually true; the researcher claims that there is a causal relationship between variable A and variable B when, in fact, there is not) and type II errors (the experimenter fails to reject

the null hypothesis; the researcher claims that there is no causal relationship when, in fact, there is one). Both kinds of error have calculated confidence intervals that indicate the certainty of the conclusion. Attempts to decrease one type of error result in an increased likelihood of making the other type of error.

Qualitative research is grounded in a philosophical tradition that is broadly interpretivist; the focus is on how the social world is interpreted, understood, experienced, or produced. It is based on methods of data generation and collection that are flexible and sensitive to the social context within which the data are produced and on methods of analysis and explanation building that involve understandings of complexity, detail, and context.

Separating Science From "Junk Science"—Call for Empiricism and Rational Explanation

Bert Black, Counsel, Diamond McCarthy Taylor and Finley, said the so-called scientific method actually refers to a number of different concepts and methods. Trial judges have been asked to determine the reliability of scientific methods used in expert evidence and the validity of conclusions as related to the matter at trial. According to the *Kumho Tire* decision, standards used in expert evidence should match those "accepted in the field" and possibly would pertain to such technical experts as police as well as scientists.[6]

Newtonian scientific thinking began with ideas of mechanistic materialism, the "billiard ball" idea of the universe, and events that were certain or predictable. Empiricism has dominated scientific thinking since that period, as researchers learned by observing patterns, collecting data, and organizing information. Explanation of results, however, has been given less consideration or guidance in this course of development. Later, quantum mechanics

counteracted the idea that "certainty" was necessarily available. The theory of relativity then altered the idea that time could be held as a constant.

In recent psychological studies, theory has been shown to affect perception (what a person thinks will affect what he or she perceives). Finally, Mr. Black quoted Karl Popper, who said theories are never absolutely proven, only well corroborated by attempts at falsification. Science itself is only a "constant dialogue between explanation and experiment."

Normal scientific discourse would incorporate empirical support, rational explanation, and some fit with other concepts of science. However, science works within paradigms that change, and any set of data can support different theories. Some key scientific experiments never used the null hypothesis approach, such as Benjamin Franklin's discovery of electricity or Fleming's discovery of penicillin, which was established by one overwhelmingly successful test application of the drug.

For lawyers, the "explanation side" of science is important. The famous physicist Wolfgang Pauli, once reviewing a proposed scientific document, said that it "wasn't good enough to be wrong." This kind of pseudoscience is that which *Daubert* means to exclude from court. Testing and falsifiability relate to the quality of empirical support behind a proposed idea, and peer review indicates the current range of scientific discourse on that topic. Error rates are applicable to specific techniques used in a process. All of these, for the decider of fact, are intended as descriptive and not *prescriptive.* In day-to-day forensic science activity, error rates for specific techniques may be an important consideration because the science of DNA analysis is not at issue. False positives, which could mean wrongful conviction, are most highly undesirable.

Trajectory of Forensic Sciences

Victor W. Weedn, M.D., J.D., Director of Biotechnology and Health Initiatives, Principal Research Scientist, Carnegie Mellon University, described the history and direction of the forensic sciences. Most convictions are still based on self-confessions and eyewitness testimony. Early forensic science grew out of legal medicine. The first recorded forensic autopsy was performed in A.D. 1302, although there is reference to medicolegal autopsies much earlier. Toxicology also found early application, as poisoning was a common means of homicide. With the Industrial Revolution came the rise of police forces and investigation techniques. The first crime laboratories were established in Europe during the middle to late 19th century. Microscopy was a basis for much of this "police science." In the United States, the Lindbergh baby kidnapping case gave an impetus to the establishment of the FBI laboratory. However, the most significant changes were brought about by the Law Enforcement Assistance Administration (LEAA), which provided block grant programs to States and local jurisdictions during the 1970s. Many crime laboratories and university programs were established at that time. The second half of the 20th century witnessed much of the professionalization of the forensic scientific disciplines. Forensic organizations were expanded and others were established, Scientific Working Groups and Technical Working Groups were impaneled, standards and guidelines were developed, and certification and accreditation programs began. DNA technology was introduced and led to increased recognition and funding as well as to broad philosophic paradigm shifts within the forensic sciences. Particularly, the use of computerized databases has transformed forensic laboratories into investigatory tools, identifying suspects without the need for conventional police investigation. Progress in the forensic sciences has also

meant growing intrusiveness of law enforcement, invasion of privacy, cynicism about the justice system, and ever more regulation. Forensic laboratories are becoming more independent, but they still need a great deal of government support to deliver the promised impact against crime.

The Gatekeeper Role: Judicial Management of Expert Evidence

Judge André Davis, U.S. District Court, Baltimore, Maryland, said the revolutions occurring in biology and technology have made a significant departure from the past. Judges today have greatly increased adjudicatory responsibilities. Judges must remain skeptical of assumptions that scientific advances will, with conscious and deliberate shaping, protect values of equality and fairness. The privacy implications of technological advances are large, and a wide range of protections developed after the 1964 civil rights legislation are at stake. Many legislative initiatives worked against discrimination, but often they were not well defined. The scientific support for genetic differences among people, a good example, creates a challenge to the protection of their equality and fair treatment under the law. A new and foreboding responsibility has been thrown on judges in reference to specialized and scientific evidence.

Judge Davis mentioned the value of court-appointed experts. He said he recently received a challenge to fingerprint evidence in one of the government's current cases. He told the group that he would be looking to AAAS for help. Judges have an independent duty to prepare for their assigned roles and are usually eager to do this. Constituents, advocates, and especially jurors are entitled to expect guidance from judges, he said.

He said he considered that judges had two options: seek extrajudicial educational

opportunities for heightened scientific literacy ("tutoring" in a pertinent area) or rely on court-appointed experts. Judge Davis said he would like to see both of these, especially greater use of court-appointed experts. He views the gate-keeping role as consistent with the fifth and seventh amendments to the U.S. Constitution.

In response to an audience question on properly defending indigents, Judge Davis said judges must do everything they can to see that the field is evenly balanced. Juries have to be advised to look for "unexamined assumptions" and the reliability of the evidence presented. The judge and jurors members are all playing roles to discover the truth, he said, and the evidence is supposed to assist us.

Mr. Fisher noted that use of expert witnesses brings up many new support issues. How can funding be developed for this "academic component" in trial practice? Crime laboratories are often significantly underfunded. Perhaps nonprofit academic research could begin to answer fundamental questions on "reliable demonstration" and acceptable error rates.

Forensic Fraud: Who's Taking the Fall? Roundtable

Sheri H. Mecklenburg and **Michael P. Monahan,** both Assistant Corporation Counsels, Department of Law, City of Chicago, discussed the implications of convictions overturned due to DNA exclusions. Their department has been examining cases that were overturned on the basis of forensics, preparing a critique of the process, and trying to see where improvements can be made.

Biological matter left at a crime scene is not necessarily from the perpetrator. Errors may occur in the handling and use of DNA evidence, eyewitness testimony, and other forensic tests. Serology in the 1980s was regarded with great confidence, but later knowledge showed it to be very inexact. What changes in the future may change legal conclusions drawn on the basis of today's knowledge? Will the original trials be affected? Civil lawsuits often follow as the released inmate looks for a source of blame for his or her imprisonment.

Law enforcement must give more attention to the collection of evidence. Some departments do not even use gloves. In some publicized cases, witnesses have been coerced or other faulty actions have occurred. Under the civil court standard, this can appear to be like a vendetta. If such a case is overturned, a defendant (who may have plea-bargained to a lesser offense) may claim to have been coerced by law enforcement. The prosecutor and crime laboratory could be sued as coconspirators. On the criminal defense side, attorneys are often targeted for ineffective assistance of counsel when possibly newly available scientific tests are performed postconviction.

Scientists are expected to make no mistakes in this society. They are often brought into the litigation by an employer or organization with "deep pockets." To establish a claim of forensic fraud, however, there must be "deceitful practice with intent to deprive another of a right or valuables." It is important to establish review processes for systemic problems in the public arena. Police and other government agencies are typically anxious to avoid blame. Mr. Fisher said that, from the perspective of a large laboratory operation, employees will occasionally do something forgetful or "stupid." If the act is reprehensible (such as falsifying evidence), they have to be dealt with seriously. But the challenge is to find an appropriate forum for less grievous systemic issues. Police, prosecutors, and courts have ignored

some legitimate claims of problems from the defense bar. Failure to use a test that did not previously exist, however, is quite different from failing to perform required work. If people use such terms as "unethical conduct" and "fraud" in a cavalier fashion, the perceptions of jurors will be tainted. This also may prevent juror and judicial trust of professional opinions, productive scientific dissension, and professional relations between laboratories.

Ms. Mecklenburg said her project wanted to look at whether a scientist can be held culpable for results that were inconclusive. Currently, about 9,000 forensic science laboratories are needed to eliminate the national forensic evidence backlog. The laboratory scientist is not more to blame for an erroneous conviction than the prosecutor, but the public considers crime laboratory staff to be scientists. The jury will hold them equally responsible whether they hold a Ph.D. or a professional technical license. A confidential laboratory employee review process with protected peer review is needed. During case processing, better communication between prosecutor and judge could help, but both parties need equal access to the scientist witnesses.

Open Discussion

Fingerprint and firearms examination are not subjects for Ph.D. degrees, noted Dr. Fisher. But these are given as expert evidence topics and are under rules of evidence. Methodologies for firearms examination, handwriting, or fingerprinting ought to be reliable, even though these experts are not "traditional scientists." Dr. Apple said that all scientific societies (forensic sciences, too) have standards of ethics. If, however, no protocol existed to do a certain test, the person missed doing that test, and the test would have exonerated the defendant, this would be improper use of the term "fraud."

A participant asked about the scope of the problem of forensic fraud. It is incorrect, the person noted, to assume that jurors consider DNA infallible. In a set of interviews with prospective jurors in King County, Washington, only about 1 in 15 thought that DNA was infallible.

Professor Risinger commented that there could be culpable misconduct if the scientist/technician is swept up in team and role planning with investigators and prosecutors. They receive too much "extra-domain" knowledge. The situation requires adoption of blinding protocols in the laboratories to insulate persons from bias and remove incentives for bias. Even careful practitioners cannot always "will themselves to resist" non-domain-specific knowledge; it has an irresistible effect. Another participant agreed that in U.S. jurisdictions, the scientists "play too closely" with law enforcement.

Friday, October 5

Court-Appointed Experts and Advisors: Panel

Margaret A. Berger, Suzanne J. and Norman Miles Professor of Law, Brooklyn Law School, New York, said this year was the centennial of an article of Judge Learned Hand from 1901: "Historical and Practical Considerations Regarding Expert Testimony."[7] In the last decade, expert testimony has taken on a new importance, and it would be helpful for courts to eliminate some of the problems in using this kind of testimony and make more court-appointed technical advisors available.

The Federal framework for using court-appointed experts, Rule 706, applies in both civil and criminal cases. The court can appoint an expert on its own motion, or the party showing cause can submit nominations from which the court selects an expert. This rule of evidence contemplated that court-appointed experts would testify at trial, but experts now are used more and more to help the court make preliminary decisions about what evidence to admit at trial and who should testify. In criminal cases, Federal courts often have funds for these procedures, but State courts may have fiscal problems. State courts have interpreted constitutional obligations differently in many situations as to when indigents must be provided with experts. Rule 706 has been upheld to give power to judges to appoint experts to gain additional information. Such new situations have no procedures spelled out and raise many issues. Should parties have access to such experts? To what extent may the judge communicate directly with the experts? The speakers today offer much experience with scientific and technical evidence at trial and selection of court-appointed experts.

Court Experts

Ronald S. Reinstein, Judge, Superior Court of Arizona, told the group about three kinds of experts he often appointed. One is an expert in "custody evaluation"; for example, in family court, each party can submit three names and strike one from the other side's list. Second, risk assessment experts may be used for violent crimes, such as sexual assault. Third, experts may be appointed in scientific areas, such as groundwater evaluations for an environmental case, computer issues, or admissibility of testing methods for evidence involving DNA analysis.

In 1995 Judge Reinstein was asked to conduct a *Frye* hearing (admissibility of evidence) for 20 consolidated pending cases involving the use of DNA analysis. Both sides presented experts on scientific issues, population genetics, and laboratory testing methods (restriction fragment length polymorphism). Two years later, attorneys came back with another consolidated hearing for 15 to 20 cases concerning polymerase chain reaction testing for DNA analysis. In the hearings, experts often said opposing things. When another hearing for a DNA testing method was considered, Judge Reinstein said he

thought additional *Frye* hearings were unnecessary and asked the attorneys if they would agree to a neutral court expert selected from a list the court would put together. They agreed to this, and soon many judges throughout Arizona were adopting Judge Reinstein's findings concerning DNA testing. Both sides save money this way, and there are no accusations of "hired guns." Judge Reinstein has advised counsel to watch carefully for case-specific issues, but usually the admissibility findings are accepted by counsel. All agreed that the results are good for circulating information and sharing or questioning positions.

An existing evidence rule tells the judge to obtain permission of the parties before speaking to an expert in advance of proceedings. Judge Reinstein recommended changing this canon because for certain types of issues (for example, those requiring computer expertise) the judge will have "no clue." The court needs an opportunity to speak to the expert or to call an adviser, such as a local university professor, concerning basic understandings in such cases.

This is an example of a weakness in the adversarial system; opposite conclusions can be reached with the same set of facts. Some concern exists that the jury will give too much credibility to the court's expert when a case goes to trial. However, jurors do not need to be advised that the court has chosen this expert. A trial is supposed to be a search for truth, and whatever a judge can do to promote this, including appointment of experts, is productive and should be considered.

Being a Court-Appointed Expert

Rebecca Klemm, President and Technical Director, Klemm Analysis Group, Inc., said the work of an expert in a large civil case is both complicated and exciting. Especially in a Rule 706 situation, no protocols exist for how the expert should assist the decisionmaker. Dr. Klemm does independent statistical analysis, reviewing reports and testimony of opposing experts and setting out the data in a manner that shows the impact of different scenarios (or questions raised). She is usually brought into a case by both parties, jointly appointed. She worked recently in a large employment discrimination case during the staging process to help the court determine when it should go forward with specific issues. As the court-appointed expert, she never spoke directly with attorneys of either side, only with her own work team and the court. Periodically, however, all groups met and shared questions. At the end, Dr. Klemm gave an extensive report and presentation and accepted questions from both sides about her analysis.

She has reported in medical matters concerning the effects of lost wages, medical care, and other impacts. She may take a group of expert reports, annotate them, and display them in tables showing the positive or negative cost impacts suggested in the material. In a major telecommunications case for which she consulted, the conflict involved parity of service and opening a market for services. She reviewed long reports by diverse experts, compared them as to detail in the generated formulas, and explained what cost aspects would matter in particular future situations.

Dr. Klemm said her analytical process is very open. She feels the public gains from this work, and her finished report is in the public domain so people can call about it and receive clarification. She said key lessons learned in her work as an expert adviser include the following:

- Keep in mind that the expert's role is that of adviser, not decisionmaker.

- Remember that even the "very smart judge" needs help to determine relevant

and salient questions, whether a particular "expert" is needed or not.

■ Present material in a manner that is easy for the decisionmaker and write plainly.

■ Consider what kind of "domain expertise" is needed, depending on the types of questions.

■ Provide details of selection procedures, show credibility, alleviate biased comments, and use tabular comparisons.

■ Show the best aspects of each expert's work.

Dr. Klemm described herself as a methodologist or statistician. She plans good ways to make comparisons. Fairness, equity, and parity are essentially the same topic in different domains. Areas to be addressed, independent of the actual domain questions, include database issues, modeling estimation, and experimental design. In situations where she has been jointly appointed, Dr. Klemm sets up audits for certain decisions in a legal proceeding. She has developed a protocol for questions that arise that has been used in many cases to avoid a trial.

Registry of Independent Scientific and Technical Advisors: A Judicial and ADR Resource

Corinne Anderson Houpt, Registry Director, Private Adjudication Center (PAC), Duke University, described the Registry of Independent Scientific and Technical Advisors of Duke University and the Private Adjudication Center, a nonprofit subsidiary of Duke. Sometimes judges or alternative dispute resolution (ADR) practitioners need specific, quick answers to technical questions. The registry was established about 2 years ago to provide a list of well-qualified experts, recruited from a variety of fields, who would be

willing to advise courts. At present, about 50 technical advisers participate in the registry, mostly from the clinical/medical fields (genetics, clinical, toxicology) and environmental disciplines.

The process of recruiting the registrants has been worked out using a network of deans and department chairs of various universities. Good experts are often identified promptly, and PAC has attempted to enlist persons who are also good communicators. Registrants adhere to a code of conduct designed to ensure public confidence. PAC helps to avoid conflicts of interest by checking funding support information and potentially conflicting commitments. They avoid including any "adversarial" experts (i.e., those who regularly testify for the same side in court proceedings). The experts receive reasonable reimbursement for time spent but not a large amount. They undertake this work as a public service for courts, arbitrators, government agencies, or parties who agree to an independent opinion.

Although the project is still somewhat experimental, experts already have been referred for eight court cases involving age discrimination, environmental science (two cases), economic expertise, and criminal law (DNA identification) and have worked on several facilitations or mediations. All communications are channeled through the registry; the judge does not speak to the expert directly. This gives greater confidence to the lawyers and parties involved. The standard service includes a written opinion responding to questions framed by the judge. In an environmental cleanup case, the community also received help in evaluating cleanup options. Depending on how the expert is appointed, the expert may be deposed or may testify. The registry is a work in progress and intends to respond to practical needs of the courts. It may also simply provide a "nonadversarial read" on a scientific or technical case issue.

Federal Courts' Reluctant Embrace of Appointed Experts

Joe S. Cecil, Senior Research Associate, Federal Judicial Center, said judges have certain real difficulties in using a court-appointed expert in the context of party-presented evidence. In a recent study of about 100 Federal judges, more than half who had used a court-appointed expert said they did it only one time. Most of the judges who had not used an expert said they would use one if they felt the adversarial process was failing and did not have the information to make a reasonable decision.

Timing of the appointment often forms a practical stumbling block. The judge may not realize until too late that he or she is "in trouble" with respect to certain information. On the eve of trial, after the end of discovery, the parties are reluctant to pay for an opinion that may hurt their case. The problem of finding an appropriate expert has improved in recent years, with the efforts of Duke University's Registry and the American Association for the Advancement of Science (AAAS) CASE project. However, another significant problem involves ex parte communications by the judge. The judge may be strongly inclined to discuss certain issues with the expert, but the parties may oppose this. The judge can, as an alternative, arrange for the parties to sit in and have communication on the record.

In 56 of the 58 cases studied in which the court reached a decision on the merits after receiving an expert's opinion, the court followed the expert's advice. This proportion gives one pause about the power given to such an expert.

Using a panel of neutral experts (as done in the silicone gel breast implant cases) can be helpful for very large, complex cases, especially when limited research has been conducted in the area. In that example, Judge Sam Pointer decided to appoint a neutral panel representing rheumatology, epidemiology, medical toxicology, and other disciplines. It is hard to attract the right people to serve on this kind of panel for the court. The professional rewards are limited, and colleagues in many fields even look unfavorably at such service. A judge who departs from the adversarial process must be willing to "put up with the storm" that grows from the significant public policy considerations.

Open Discussion

Professor Berger noted that the group was actually looking at several different models of expert advice: tutorial for the judge, a responder to parties' questions, and a panel of experts to render a decision. At what point is the expert best utilized? How should information be integrated? Should the jury be told that the court appointed the expert, or does that give too much prominence to the expert's opinion? What works with these difficult practical issues?

Dr. Cecil said the court can require the parties to nominate the expert or the judge can make the appointment from the bench. The CASE project of AAAS grew from discomfort among members of the scientific community who were disturbed by the negative view of colleagues who testified in court. The result was that some of the best scientists refused to testify because doing so would not further their professional goals. Professor Berger agreed that many experts are "horrified" by the experience of being cross-examined, but lawyers know the experts have to be prepared for the questions they are likely to receive. One participant suggested that courts could attract more scholarly help by offering research opportunities within the judicial system as a kind of "quid pro quo."

Dr. Klemm noted that the professional communities' values are changing. Scientists have been public speakers

much less often than lawyers. Cross-examination puts specialized knowledge on the line and makes the person see what he or she does not know as a scientist. It is valuable, however, for society.

Law Enforcement Personnel Testifying as Experts: Panel

Carole E. Chaski, Visiting Fellow, Office of Science and Technology, National Institute of Justice, introduced and moderated the panel, replacing Professor James Fyfe, who was unable to attend.

Challenges Facing Law Enforcement Personnel Testifying as Experts

Carol Henderson, Professor of Law, Shepard Broad Law Center, Nova Southeastern University, said that law enforcement professionals are taking the witness stand as expert witnesses more and more often and they face many challenges. The fields in which they are frequently called include drug recognition, criminal modus operandi, digital evidence, police practices, profiling, and accident procedures. Experts (serologists, pathologists, fingerprint experts, forensic dentists, toxicologists, and professionals in other fields) have been accused of fraud and negligence and, sometimes, "systematic corruption," so investigation of their qualifications is becoming increasingly thorough.

It is important for attorneys to research the experts chosen and look at their qualifications, including education, employment, publications, awards, and professional associations. Although lawyers are increasingly using computer resources and Internet case research, only a small percentage of law schools teach courses in science and the law. According to a congressional study, more than 500,000 Americans are employed on the basis of

fraudulent credentials. For example, some "board certifications" are given simply on the basis of points for attending meetings.[8] Attorneys need to become educated regarding the bases on which various organizations grant board certification. Transcript summaries posted by WestLaw, NetCourt, and Medical Malpractice Expert Witnesses also are invaluable in weeding out witnesses with fraudulent credentials.

Law enforcement witnesses often expect the jury to view them positively, but recent trends do not necessarily support this view. In a survey (Saks) of jurors' perceptions of trustworthiness, nurses, physicians, chemists, and firearms specialists ranked high, and polygraph specialists, police, and handwriting experts ranked low. Among the younger generation, many lack trust toward authority figures and react favorably toward competent, automated presentations in court (PowerPoint, video). Confidence in crime laboratories has been shaken since the Joyce Gilchrist scandal; about 84 percent of those surveyed thought that Jeffrey Todd Pierce, who was wrongly incarcerated as a result of Gilchrist's forensic testimony, should be financially compensated.

Judges, too, have become more distrustful of experts and have excluded their testimony more frequently. Neither credentials nor peer review have alleviated judicial concern about bias. Judges' understanding of current principles in the psychology field is mixed. In the Kovera study, 17 percent of the judges "admitted into evidence" statistically flawed studies.[9] The trend toward accreditation in the forensic field is favorable, but the justice system cannot ignore the popular culture and mindset. Poor communication techniques, lack of visual aids, and unclear language form major obstacles to juror receptivity.

Training workshops for technical staff serving as expert witnesses are very helpful. Improved communication skills

through the use of moot court, speakers groups, review of transcripts, and studies about communication can help prepare for appearances in the court setting. The expert should never exceed his or her qualification and expertise. Presenters need to know how to use the current technology, including enhanced photos, animation, and virtual reality. These techniques are more likely to attract the attention of today's jurors.

Police as Expert Witnesses

Philip J. Cline, Chief of Detectives, Chicago Police Department, told the attendees that police are often called to testify in both civil and criminal trials. A police officer qualified in gang crime, for example, may testify on the significance of words, the evidence of motivation in a gang shooting, or the background of one gang's relationship to another. Police also testify on estimation of the value of street narcotics to indicate the amount of profit involved in a criminal transaction. The qualifications and experience of the individual police expert must be looked at carefully, considering education, years of experience in a particular occupational area, and specialized training. Chief Cline said, for example, that he would not himself qualify as an expert in matters relating to traffic because he had only about 10 traffic cases in his 30 years of policing.

Police officers have to keep up their professional development, expertise, and education; time on the job is no longer enough. DNA has been a "double-edged sword." Police want to "put the right people in jail," but they have to be properly prepared for new kinds of investigation. Quite a few recent cases have been seriously unfavorable to police credibility, such as the "Chicago 7" officers who were convicted of operating a drug ring. Chief Cline said efforts to use new technology on the street, videotaping over a

6-month period to show citizens how the police were working, began to swing public opinion back in favor of the Chicago police.

Law Enforcement Personnel as Experts

Bruce M. Lyons, attorney, Lyons and Sanders, Fort Lauderdale, Florida, said the use of experts has been important to defense attorneys and has strongly affected decisions in court. The impartiality of a paid expert is always a concern. His law practice has benefited from meetings on new technologies and calls made to pathologists and experts such as Professor Imwinkelreid or Professor Henderson, who were willing to engage in dialogue. In postcase analyses, 70 percent of judges and lawyers found scientific evidence more credible and scientific experts more persuasive than in the past. The expert's role has to include helping the court to understand the evidence and how it is being applied. Testifying as an expert, a law enforcement officer who says that one ounce of cocaine is more than "for personal use only" may have a devastating effect on the defense. Defense counsel will have to look into other statements of the officer to check consistency.

Police experts are frequently used to fill in possible explanations for missing evidence (e.g., to explain why no fingerprints were found on a large quantity of marijuana brought into the courtroom as evidence). The judge and jury may need to be educated about the evidence. An "old fashioned" judge might be prepared to admit objectionable handwriting evidence just because he "usually does." Some police witnesses are admitted to testify as lay witnesses, but the jury still treats them as experts. This is especially true with relationship to fingerprints, handwriting, modus operandi/criminal profiles, crime scene reconstruction, and narcotics values.

Police need to be careful now about this kind of expert testimony. In the past, police have often described general practices of criminals in court, such as how a drug courier drives on the highway, following trucks slowly, etc. Such profiles are sometimes no longer admitted. Police often give evidence on street slang, but this can also be tricky and is sometimes rejected on appeal. The court can also use jury instruction to point out to jurors that some claims are technical rather than scientific and to try to limit the impact of the witness's testimony on the jury.

Eyewitness Evidence: Panel

Moderator **Sandra I. Rothenberg,** Judge, Colorado Court of Appeals, noted that she had been one of the State trial judges tested during the survey referred to earlier in the conference. She said, "Let us not compound errors by punishing innocent people." Eyewitness testimony will always be important evidence, and new techniques for memory, using cognitive psychology, can improve eyewitness identification. She asked everyone to think about how eyewitness evidence affected the work of the justice system and courts.

Getting the Lawyers to Listen

James M. Doyle, Attorney, Carney and Bassil, Boston, Massachusetts, said eyewitness evidence is the oldest form of evidence, and it has been an old battle to get social sciences into the court. He referred to a 15-year-old reference book, *Eyewitness Testimony, Civil and Criminal,* which is still not often read.[10] Hugo Munsterberg studied applied psychology and challenged the idea that humans are equipped with permanent memory faculty (contradicting the idea that a person might forget but would not remember the wrong person). A response in the *Illinois Law Review* examined 149 articles on eyewitness psychology.[11] It accused Munsterberg's approach of

"libeling the legal profession" and casting all questions of science as "only for the expert." The discussion highlighted the differences in the types of knowledge sought by lawyers in contrast to psychologists. The scientist's focus of interest is the probability that the result is correct (e.g., will it be correct 8 out of 10 times?). The lawyer, however, wants to know whether this *specific* case is "one of the eight" (most frequently occurring) or "one of the two?" In other words, the different professions are looking for a different kind of "reliability," one statistical and one diagnostic.

Scientific evidence is more probabilistic than diagnostic, dealing with what is "normal." Testimony about what is normal "destabilizes" a great deal of existing case law. In one year, about 75,000 cases turn on eyewitness testimony. After DNA evidence became available, it was found that 82 percent of wrongful convictions were based on eyewitness testimony. For judges, the "only for the expert" way of thinking is too debilitating. They become preoccupied with the jurors' confidence levels and concerned that they will be overwhelmed by the material. It would be more useful to develop the confidence levels of the judge and attorneys, who often know as little as the jury members. The NIJ guidelines on eyewitness testimony bring forward preventive measures that can be taken. The key insight is that diagnostic work (not probabilistic) will have to be done in the courtroom. Memory evidence is trace evidence that may be subject to contamination. When or how evidence might have been contaminated must be considered.

Improving Juror Decisionmaking With Scientific Research on Eyewitness Memory

Ronald P. Fisher, Professor, Department of Psychology, Florida International University, said that police receive little

training in collecting eyewitness evidence. Often, they simply have a list of "facts" that they try to elicit, but the necessary skills for effectively interviewing witnesses are difficult to acquire.

A typical law enforcement interviewer often interrupts the witness and takes a very active role rather than allowing the person to volunteer information. This does little to facilitate the witness's memory. Leading questions may actually distort memory considerably. The device called the cognitive interview (developed in journalism and other fields) recognizes the social interaction between the interviewer and witness and is based on psychological factors (memory, social dynamics, communication). "Cognition" implies an effort to recreate the original context by asking the witness evocative questions. At any point in time, some facts may be more accessible to the witness than others (at least temporarily).

The interviewer should not cause the witness to use mental resources for "something else" (side topics). Social dynamics most usefully follow the pattern that the witness is "knowledgeable" and the interviewer is "curious." Witnesses may expect the interviewing officer to "act the star" and do most of the talking (they should be disabused of this idea). Police departments have noted that some personnel are more effective interviewers naturally, and departments can test to see which recruits are well suited for this.

In a laboratory study of the cognitive interview, it was found that 30 to 70 percent more information can be collected using this technique. Accuracy, however, is only slightly higher (85 percent compared with 82 percent). In one study, trained high school students using the cognitive interview technique collected almost 100 percent more information than police with 15 years of experience. Metastudies are

being conducted in England, Canada, and Germany. The United Kingdom trains all police officers in the cognitive interview technique. Additional benefits of this kind of interviewing include better retention of (more satisfied) witnesses, increased confessions, and greater closure of cases.

Eyewitness Identification: Scientific Research and Application

Gary L. Wells, Distinguished Professor, Department of Psychology, Iowa State University, spoke about the collection of eyewitness evidence before it gets to court. After reviewing DNA-related exonerations, he started studying a scientific method for approaching eyewitnesses, using a trace evidence metaphor. He said the most pressing need in this area is for greater use of blind testing in law enforcement (police department lineups, etc.). More than 80 percent of initial identifications come from photographs (which could be out of date), and wrong identifications can be made.

Dr. Wells recommended focusing on system variables, the elements most easily controlled, to increase accuracy. The perpetrator has left an image behind in the brain of one or more eyewitnesses. How this evidence is handled matters because, once it is contaminated, it is no longer trustworthy. A natural tendency exists to make relative judgments and "shift the choice to the closest next best" rather than to make no choice at all. For this reason, it is important to have no more than one suspect in any lineup and to select fillers who fit the description closely but are known innocents. Observations of the relative judgment problem led to the development of sequential procedures for identifying suspects. This procedure makes the witness "dig deeper" into memory and reduces the rate of mistaken identification.

Blind testing is the most important principle to implement. The law enforcement tester must not know who the suspect is (or he/she may inadvertently communicate this to the witness). Strong verbal and nonverbal influences may affect the viewer through the lineup administrator. It is an interactive situation (e.g., the administrator might hint, "Be sure to look at everyone . . ." or "Did you pause at number 3?"). Body language displaying interest in a particular person in the lineup will also "give clues" to the witnesses.

Changes are emerging in some jurisdictions. New Jersey has fully adopted the NIJ guide, including blind testing and sequential procedures. New York is exploring the use of the guide, as are Iowa, New Mexico, and Hawaii.

Science and Eyewitness Evidence, Challenge for Law Enforcement

Mark Larson, Chief Criminal Deputy, Criminal Division, Office of the Prosecuting Attorney, Seattle, Washington, said he had participated in a lot of multidisciplinary work with police officers. Police are the "forensic specialists" for eyewitness evidence, but they do not perceive work with eyewitnesses as a science. Too often, they are simply taught to follow a protocol and fill out forms. An opportunity exists here to apply scientific theory and principles to the highly dynamic fact patterns faced in the field. Mr. Larson said he hoped departments were working with the NIJ guide in their training protocols and practice. Legislative or political mandates (as in New Jersey) could bring about more careful handling of eyewitness testimony.

Mr. Larson said he would like to encourage collaboration between law enforcement and social science to bring different professional communities closer together. The National Center for Eyewitness Evidence is a good example of what can be done to aid training, guide research, inform police, and improve work against crime.

Open Discussion

One participant asked whether other protocols for trace evidence could be applied to eyewitness evidence. Although nothing is wrong with using guidebooks, Mr. Larson noted, law enforcement has to use its judgment concerning situations in the field. Another attendee said State law in California does have qualifiers or prerequisite conditions for conducting one-on-one questioning.

Dr. Wells responded to a question on controlling for cross-racial identifications or ethnic group variability. Research shows consistently that cross-racial identification is more difficult than identification within race; this is true between any variation of racial group and is not really a system factor. A useful system intervention might be to use additional fillers in a lineup and to be careful to use known innocents.

Another person asked whether officers using the NIJ guide, in New Jersey for example, would continue to show photographs in the sequential methods after a person had been selected. Dr. Wells said photographs can be shown both ways, but he considered it to be better to set the procedure in advance rather than to stop when the person picks someone. Innocent "fillers" should resemble the perpetrator and the suspect but not be "clones."

Judge Rothenburg asked the panel whether it was good for judges to instruct the jury about dating of photos and similar areas of caution. She knew of one situation in Los Angeles in which jail inmates had simply been allowed to pick the lineup fillers. Dr. Fisher noted that, in mock jury cases, better results occurred when eyewitness guidelines were described for the jurors in advance.

Medical Records as Legal Evidence of Domestic Violence

V. Pualani Enos, Assistant Clinical Professor, Domestic Violence Institute, School of Law, Northeastern University, said she realized how necessary factual medical documentation was for legal evidence in abuse proceedings when she noticed that professionals of all disciplines could not help wishing to avoid the uncomfortable topic of domestic abuse. Her study revealed some important shortcomings of current methods of documenting medical charts when they may become legal evidence. Dr. Enos found common goals among experienced health care workers and attorneys who are willing to work together to increase understanding in abuse cases.

Change in the medical system's response to domestic violence has been slower than the dramatic changes in the justice system, particularly in view of legislation during the last 5 years. As the consequences to batterers become more severe, the standards of proof that domestic violence has occurred are also rising. People in the health care setting are the most trusted by traumatized victims and are most likely to have observations and statements from the victim that cannot be located elsewhere. History, mechanism of injury, patterns, and consistency of statements can often be found in health care documentation. Attorneys have great difficulty obtaining access to such records. Lawyers often finally give up on getting this type of evidence admitted in court. People consider such evidence to be important to criminal cases, but it also may be important for immigration, housing, or even special education.

The goal of the research was to encourage production of more accurate and comprehensive medical records to make available information relating to diagnosis and treatment of domestic abuse that might be needed in court. Health care providers should understand that a failure to *document* domestic violence completely when treating it will almost always convey a legal advantage to the abuser and constitutes poor preventive medicine. Dr. Enos studied cases in the context of evidentiary rules, some usually loosely applied and others usually strictly applied. Study subjects were recruited in the hospital and in court. In the course of 772 visits to the hospital, the research reviewed 96 medical charts of 86 abused women that presented evidence of physical, sexual, and emotional abuse as well as stalking. The records came out of primary care, obstetric/gynecological care, and the emergency room. In 19 percent of the records, the perpetrators were indicated by relationship, but they were very seldom noted by name. About 20 percent gave some kind of psychiatric diagnosis but did not mention domestic violence. Nurses provided most of the documentation found in medical records.

Health care workers are trained to be sparse in their descriptions. In an effort to be "neutral" regarding abuse situations, medical personnel may use language that could hurt the injured person's legal case. If providers just use "patient states" or place quotes around a patient's comments in documentation, these could then be admitted in court as "excited utterance" under evidentiary rules. An excited utterance may be admitted without the victim having to testify, if it is explicit about the event. As many as one-third of the notes from doctors or nurses were illegible, making them unusable for the victim's court case. In most cases, the health record notes were also too vague. Records would be improved by using photographs routinely, noting the time and date of an injury, and describing the patient's demeanor.

When a case involves "injury by a stranger," the record tends to include a lot more detail than when a partner is involved. The legal and medical professions hold many misperceptions of each other's roles. It is particularly undesirable for medical providers to use legal-type wording, such as "alleges," "assault," or "perpetrator." The meaning of this language is very specific in a court context and can raise doubts about who created the injury. Most providers fear going to court and want to reduce any time the health care provider staff would have to be on the witness stand.

The project's effort to access records continued for more than 12 months and proved to be very expensive. Many confidential topics, such as HIV or sexual assault, were addressed. The study has been able to develop training for doctors and nurses in better documentation of abuse cases and has evaluated the incorporation of medical documentation into litigation. Dr. Enos highly recommended that legal organizations and medical records departments collaborate in areas of risk management.

Kumho Tire and the "Task at Hand"

D. Michael Risinger, Professor, School of Law, Seton Hall University, South Orange, New Jersey, said that although *Daubert* was intended to make admissibility of scientific testimony more rigorously controlled, nonscientific evidence has been handled in varied ways. *Daubert* standards were too often made into a "mechanical checklist," as courts often failed to grasp that under *Kumho Tire,* reliability of evidence cannot be judged "globally" but must be judged specifically as applied in each case. In *Kumho,* the plaintiffs' vehicle's tire blew out. They argued that the tire was defective because tread separation occurred. It was uncontested that a nondefective tire would not separate as a result of normal driving, but a certain form of tire abuse stemming from long-term underinflation was claimed. The tire failure analyst, Dennis Carlson, said that any tire subject to such abuse would show tread wear on the edges of the tire greater than in the center of the tire, a groove worn on the tire's bead (rim of wall), discoloration on the wall, and marks on the flange itself.

At issue concerning this evidence was not the use of visual inspection, but the reasonableness of using that approach for the specific (not general) question before the court. There was no general acceptance in the community of Carlson's kind of analysis; acceptance in a *Daubert* hearing cannot show reliability where the discipline itself lacks reliability. Professor Risinger supported the use of varying levels of "foundational reliability" for expert evidence when criminal guilt (liberty or life) was at stake. Proper task-at-hand analysis identifies the evidence variable specifically (e.g., a specific example of handwriting in context). The court is allowed considerable leeway for determining reasonable reliability for the task at hand and is not required to meet the same threshold in every context.

Mere experience as basis for the reliability of an expert witness is sufficient only when the person plays a "summarizational" role for the court, such as describing an industry based on many years working in that field. Professor Risinger said dependability of evidence might be a better phrase than reliability, which has not been clearly defined in Supreme Court cases. He referred to a judge's rejection of handwriting "expert" testimony in a recent questioned-document case as a good example of "case in hand" analysis. According to Professor Risinger, a judge excluded technical handwriting testimony when the context of writing styles involved printed (not cursive) script and originally non-English-speaking individuals.[12] When reliability does not depend

on the expert's experience, it would be irrational to use that experience as a basis for admitting evidence.

Recent Defense Challenges to Forensic DNA Evidence

William C. Thompson, Professor, Department of Criminology, Law, and Society, University of California, Irvine, told participants that he had reviewed a number of defense challenges to DNA casework in recent years. In some cases, ambiguities, limitations, or problems were not disclosed in laboratory reports, and many difficulties were related to poor scientific practices in the forensic laboratory. In particular, the failure to use blind procedures sometimes cast doubt on interpretation of forensic laboratory results. Scientists interpreting the tests are often involved in the case investigation. In addition, the technician's laboratory notes may reveal partiality for or against a suspect (e.g., ". . . want to connect this guy to the scene" or "keeps skating . . . never serves time"). From psychological studies, he said, we know that people with strong expectations as they approach ambiguous data will tend to disregard interpretations that differ from the one they expect. Expectations may lead them to resolve ambiguous data in a manner consistent with those expectations or miss or disregard evidence of problems and alternative interpretations for data.

Most attacks on DNA evidence claim that the evidence was compromised by handling/processing errors, inadvertent transfer of DNA, biased interpretation of results, or exaggerated (or misleading) statistics. Even the newer technologies can be ambiguous in some cases. Moreover, although a person's DNA is found at a crime scene, it may be unclear how it got there. Inadvertent transfer of DNA from an individual to an object or another person is possible and occurs more easily than previously thought. C. Ladd and colleagues have investigated the potential for and effects of DNA transference.[13] Even a minor contact, such as a towel touching a face, can transfer DNA in measurable amounts. Quantities as small as 15 cells can be detected in tests.

Statistical interpretations have often been biased against the accused, overstating the likelihood of a match, especially when the evidence is a mixture of different DNAs. Technicians are often willing to declare two profiles a "match" even when they do not match perfectly. Thompson presented examples from actual casework in which technicians had declared a match despite differences between the DNA profiles of the evidentiary sample and the defendant. The differences were attributed to a variety of phenomena, such as degradation, artifacts, and allelic dropout. Because the standards for declaring a match are vague and flexible, it is difficult to estimate the true likelihood that a laboratory will declare a match between profiles of different people.

Sample-handling errors have also occurred. In one rape case in Philadelphia, the laboratory mixed up the reference samples of the victim and defendant. The laboratory reported that the defendant's DNA profile was found in a vaginal swab. The defense investigation revealed that the vaginal DNA was actually that of the victim herself. False positives are particularly worrisome in "DNA dragnet" and "cold hit" cases where there is little reason (a priori) to expect a particular suspect to match. Statistical analysis shows that even a relatively low false-positive rate may significantly undermine the value of DNA evidence in such cases. And defendants in these cases face due process problems. Informing the jury of a "cold hit" in an offender database may bring up inadmissible prior crime evidence. Not informing them about it may cause the jury to overestimate the strength of the case.

Saturday, October 6

Digital Evidence—Virtual Reality in the Real-World Courtroom: Mock Trial

Introduction

Susan M. Ballou, Program Manager, Office of Law Enforcement Standards, National Institute of Standards and Technology (NIST), explained that this session was a mock trial[14] involving electronic evidence and crime scene investigation relating to computers. She noted that this was a "production" and not a real trial, intended as a demonstration of important factors in using computers and computer evidence in the justice system, a necessity that appears more and more often. Ms. Ballou introduced the "cast" of the production, consisting of judge, prosecutor, defense counsel, law enforcement investigator, State crime laboratory's criminalist, and defense expert witness. The staged case was based on an actual murder case that involved computer evidence, with names and details changed for privacy.

Trial Preparation [In Role]

The judge [played by **Judge Judith Ford,** Alameda County Superior Court, Oakland, California] called the hearing into session and told the prosecution witness to take the stand. The prosecutor [played by **Richard Murray,** Assistant U.S. Attorney, Western District of Michigan] introduced the law enforcement investigating officer in charge of computer evidence recovery. The law enforcement investigator [played

by **Terry D. Willis,** Officer-in-Charge, Computer Crime Unit, Los Angeles Police Department] said he had been called to assist police officers at the scene of a murder in which a computer, with the power on, was found beside the victim. The victim died from a knife wound in his apartment, which was located over a bar. The bar manager had found the body at 2:30 a.m. The investigator said he contacted the State crime lab's criminalist to examine the computer for clues relating to the murder.

The victim ("Davidson") was found to have a dialup Internet account, the type that uses a dynamic protocol account (which changes the computer identifier—server of the Internet service provider (ISP)—for each session). The victim's computer had the Internet relay chat (IRC) software called MIRC, a popular communication protocol that links users on chat channels, similar to AOL's Instant Messenger. Users can create their own chatrooms, and many universities allow this communication forum. The prosecutor's questions made clear that no single person or company is in charge of this kind of Internet service; a chatroom participant could be an expatriate, a foreign citizen, or a member of an occupational association using the chatroom for low-cost communication.

Using a warrant issued for the ISP, the State criminalist traced the IP address to another provider using a communication from "Hot4U" that was on the crime scene computer. The second computer had a cable modem account belonging to

the suspect, Mr. Doakes. This type of account is usually hard-cabled to a particular house, often using a television. The investigation located correspondence on Doakes' computer and unopened e-mail that indicated Doakes had purchased a knife.

The prosecution's questions brought out that suspect Doakes had used an alias in the chatroom ("SeekingLuv") and was having difficulties with his marriage. Chatroom logs had been found on victim Davidson's computer indicating conversations between victim's alias "Dawn" and "SeekingLuv." Role-playing is a common chatroom activity, sometimes using sexually explicit language. Participants can create many names, share images, and purport to be separate individuals.

The prosecutor asked the investigator about a theory of motive appearing through the chatroom logs. The investigator said the victim was suspected of using a female alias ("Dawn") to entice the suspect ("SeekingLuv") into sexually explicit conversation. Using another alias ("Capt. Dread"), the victim then threatened to show the logged conversation to Doakes' wife and employer unless he was paid $5,000. Prosecution considered that suspect Doakes later used the "Hot4U" alias with "Capt.Dread" (the blackmail message sender) to locate the victim and murder him.

The prosecution introduced several questions about the first responder at the murder scene. The officer first on the scene had looked at files on the victim's computer, although he had not been trained in computer evidence handling. He was unsure whether he had "done something" on the crime scene computer before the arrival of Mr. Willis, the investigator specializing in computer recovery. The investigator told the prosecution he had pulled the power cord from the back of the computer when he arrived at the crime scene,

rather than executing the standard shutdown procedure, which usually writes over certain current files and might obscure what was on the computer at that particular time. Law enforcement analyzed a specially created duplicate of the computer's hard drive, using CaseView software, to maintain integrity of the original evidence object. After gaining access to the suspect's computer, the investigation found some material from a newsletter he published, evidence of a recent knife purchase, and text fragments from deleted files concerning a plan to stop a blackmail.

When the prosecution finished with the witness, the judge asked defense counsel [played by **Robert S. Vance** and **Anthony Joseph,** Johnston Barton Proctor and Powell, LLP, Birmingham, Alabama] if they wished to ask questions. Defense counsel stood to make a motion to suppress evidence.

Mr. Vance said that information was obtained from suspect Doakes' computer in violation of the Cable Act and the Privacy Protection Act. The suspect's ISP should have been barred from disclosing personally identifiable information on subscribers; and defendant Doakes, as a newsletter publisher, had been denied an opportunity to contest the search warrant that gave government attorneys information from his computer (as the Cable Act would allow). He said the government would have been unaware of the knife purchase had it not improperly taken and read the suspect's unopened e-mail. Law enforcement had shown no clear search strategy, nor had they disclosed information to the magistrate about the items for which they were searching. Text fragments from Doakes' computer and the map found on that computer were obtained illegally. The prosecution failed to provide clear and convincing information that subject had engaged in any criminal activity or that the computer evidence was material to this case. Mr. Vance added that there

was no clear evidence that his client, Mr. Doakes, was involved with the murder, as 12 or 14 other chat partners could also have been suspected.

The prosecutor, Mr. Murray, conceded the facts as accurate but said that Congress had amended the Cable Act in 1992 to allow an exception when the government seeks digital evidence by court order in a criminal proceeding. He said the government considered evidence of the blackmail to be a convincing motive for the murder. Judge Ford clarified that the suppression of material on the computer was sought because of commingling of publishing products of the defendant. After listening to both arguments, Judge Ford denied the motion to suppress. She pointed out the amendment to the Cable Act and said the requirement to notify the Internet customer did not apply in a criminal investigation. She said the court saw the tension between the different statutes, but suppression of all evidence would be too drastic a measure; suppression of particular protected material could be accepted by the court.

Open Discussion [Out of Role]

The participants and audience considered whether digital evidence should be treated the same as physical evidence. The investigation considers these to be similar, especially in the absence of actual witnesses. Mr. Murray said the environment of computer evidence is changing so fast that some statutes are not in sync. Admission of evidence may be the centerpiece of a case. Excluding digital evidence, similar to discrediting physical evidence, would be a major step. Mr. Willis said he considered pulling the power cord of a Windows computer as an attempt to preserve physical evidence, something left in memory at a particular time.

The question of commingling publishing work product with other types of electronic

communication relates to a developing area of privacy law. Almost any person who puts up a Web page, said Mr. Murray, currently may be considered a "publisher." In addition, many States, including New York, have not ruled about cable-provided services, but the Privacy Act repealed the requirement that subscribers be given notice of a warrant when a criminal investigation is involved. Judge Davis added that evidence issues will often be decided by magistrate judges, who handle court orders for evidence and search warrants.

Mr. Vance reviewed recent developments in expert evidence rules in response to a question on applying changes within different State settings. In *Frye,* criminal defense had tried to introduce expert evidence regarding a crude polygraph procedure. The court, agreeing to exclude this, said that it had not "received general acceptance in the relevant scientific community." Following that, the *Daubert* decision, relating to a claim that a prescription drug caused birth defects, held that the plaintiff had no scientific medical evidence that proved a *causal link* to the damages. The Supreme Court later said that *Frye* was too restrictive and formulated a standard that the expert evidence should be "sufficiently relevant and reliable to warrant admission." This has emphasized the role of district courts as gatekeepers for expert evidence, an effort to prevent "pseudoscience" from influencing juries.

Expert witness evidence has become increasingly important in both the Federal and State judicial systems, Mr. Vance continued. In recent years, the laws have changed again in relation to how courts must handle admissibility. The distinction between admissibility in State and Federal courts is based primarily on procedural aspects. State courts tend to be more "loose" about admitting evidence, but the Federal district courts repeatedly have been charged to be gatekeepers and to scrutinize evidence carefully. Concern

about proper methodology has been great since the case in which an Oklahoma prisoner (Jeffrey Todd Pierce) was released following a wrongful rape conviction. That State subsequently opened investigations into numerous cases involving testimony of the same crime laboratory chemist (Joyce Gilchrist).

The judge must determine whether the expert evidence will assist the "trier of fact." Factors to consider include the following: Can the evidence methodology be tested? Has it received peer review? Is there a known or potential rate of error? The focus is not on the conclusions of the expert but on the methodology used to arrive at the conclusions in evidence. The Supreme Court has also reaffirmed an "abuse of discretion" standard in relation to lower court decisions (that is, the higher court should not use a "more stringent standard of review"). Mr. Vance noted, however, that methodology and conclusions are not always easy to separate.

A third consideration for expert witness evidence emerged through the *Kumho Tire* case. This concerned whether evidence coming from technical but nonscientific witnesses (e.g., an engineer or law enforcement officer) also came under *Daubert*. Witnesses giving technical evidence that relies on experience must be able to explain how conclusions are reached, why there is a basis for them, and how reliably they can be applied to the facts of the particular case. Mr. Vance said there is still considerable judicial uncertainty about interpreting recent Federal rules amendments applying to expert testimony. The newest advisory committee comments seem to introduce yet another change, in that they appear to impose a degree of quantitative inquiry, whether there are *sufficient* facts or data.

One participant asked for advice about using PowerPoint and similar technologies in court. Judge Ford said PowerPoint

presentations must be shown to opposing counsel and the court ahead of time as demonstrative evidence (similar to a chart) and should be consistent with foundations that have been laid. Judge Davis said his courtroom encourages these presentation technologies, and the systems department in Baltimore will tutor any lawyer who requests it. These presentations save time and the jurors like them very much. Judge Ford told participants that various materials prepared for the mock trial (written motions, search warrants, etc.) could be viewed on NIJ's Web site.

Case Presentation at Trial [In Role]

The prosecutor, Mr. Murray, called the State crime laboratory criminalist [played by **Doug Elrick,** Criminalist, Crime Laboratory, Iowa Division of Criminal Investigation] to the witness stand.

[*Aside:* Mr. Murray noted that the strategy for this case would rest very much on precision of discussion and interpretation of Federal rules. In law enforcement, there is currently no consensus and little cohesion on the idea that an agent must be an expert just to testify about computer evidence. Standards are localized and greatly varied. He recommended careful use of etiquette between opposing counsel (giving notice of expert or technical witnesses). If the government offers a witness with specialized knowledge, it should flesh out the basis of that specialized knowledge for the jury's and court's understanding. He said a trial that is lost before a jury may never get to appeal.]

The prosecutor asked Mr. Elrick about his background in forensics. Witness Elrick said he had worked in computer evidence recovery for 12 years. Previously, he was a drug chemist in the crime laboratory. He had received specialized international training in computer recovery and completed a certification course. Mr. Elrick belongs to

the International Association of Computer Information Specialists and has examined more than 1,000 computers in criminal investigations.

Mr. Murray asked witness Elrick to explain what had been done with the computers in the present murder case. The witness said an inventory of the system was prepared, showing what components were installed (sound card, network card, etc.) and what software was found on the system. The hard drive was removed and put on a forensic computer to make an exact duplicate, which would be used for further analysis while preserving the integrity of original evidence. CaseView software was used to make this accurate representation of the hard drive and to view fragments found on the drive.

The prosecutor asked Mr. Elrick if Case-View was a commercially available software program (yes), and whether Mr. Elrick had received training in that software (yes). Mr. Elrick said that he had personally been involved in the testing of the CaseView software. The law enforcement community had tested the software extensively, and it was reviewed by *Computer Security* magazine, which verified its functionality. No statistical analyses of the tests, however, were available. Mr. Elrick told the prosecutor that the software company had coded and upgraded the software on occasion to meet law enforcement requests as it was developed. Concerning interactions with operating systems, CaseView was not able to reveal source code for Microsoft operating systems and may experience difficulty finding out how file systems are laid out in that kind of system. Mr. Elrick said the field of computer forensics is trying to keep pace with changes in operating systems, including Linux and Unix.

CaseView, said Mr. Elrick, makes a snapshot of the original hard drive and develops a proprietary file for examining copied

contents of the disk. The software uses a mathematical formula known as MD5HASH to represent all data found on a specific drive. If any data are altered later, the mathematical process arrives at a different formula, revealing that something changed. MD5HASH is a validation measure also widely accepted outside law enforcement.

Using the CaseView program on the victim's (Davidson's) computer, Mr. Elrick found MIRC software (for chatrooms) and said the chat logs were prepared automatically when the MIRC program was run. Mr. Murray asked a few questions about the nature of Internet Relay Chat (IRC) and the MIRC program. Individuals connect through the Internet through their respective ISPs, and each must have the client software on his or her PC. People often choose aliases and may change their name for different chatrooms. ISPs keep minimal records of names, checking only that they are unique as session identifiers. Mr. Elrick found a primary alias on the victim's computer, "Dawn," and a secondary alias, "Capt.Dread." The chat logs are audit functions that can be toggled on or off. Logs can be created for chat events associated with a specific "named" participant, and the computer's date/time function labels each new entry, which is then automatically appended to the named log. Mr. Elrick said that the victim's computer had more than 100 log files, while the suspect's (Doakes') computer had no logs and the logging feature for the chat software was turned off.

For the jury, Mr. Elrick briefly described the layout of a hard drive, with multiple platters made of concentric circles called tracks. These are broken into pie-shaped sectors, so each track will have "sector 1," "sector 2," and so forth. Writing a file to disk must be done in clusters; three or four clusters usually make up a page. Doakes' computer used the Windows 95 operating system, in which eight sectors per cluster would be recorded. Unused

parts of a sector will contain remnants of the previous content in that sector. When a file is deleted, the first character of the name is changed so it cannot be displayed, and the File Allocation Table is told the space is available for new material. But the older material is not actually gone and may be recovered with tools like CaseView software. The user cannot control these "remains" left in the unallocated disk space. Mr. Elrick said it was similar to "throwing away the card catalog but leaving the books on the shelf." Mr. Elrick said he had found a number of logs on the victim's computer but could not say whether all online conversations had been reported in the logs.

Defense counsel, Mr. Vance, rose and asked for more details on the background of the State crime laboratory witness. Before his work for the police department, Mr. Elrick had not done any computer data recovery. His expertise was as an end user, originally with gaming software, rather than as a writer of code. Although he participated in the software's testing, he had not been involved in the development of CaseView. Mr. Elrick did not have any specific information about the analysis of CaseView by *Computer Security* magazine. He believed they tested the imaging, copying of drives, and search features but admitted his knowledge came only from the magazine article.

Mr. Vance asked him whether, in his experience, software programs often had "bugs" or errors in code that would not allow the program to work correctly. Mr. Elrick agreed that most programs went through a process in which newer versions corrected earlier problems with the program. Newer versions and upgrades of CaseView had been used prior to this trial, but Mr. Elrick said that no independent testing had been conducted on new

versions of the software (which would have been very expensive). Mr. Elrick had no knowledge of independent testing by disinterested parties, nor did he know of organizations that would do such testing to promulgate standards for the software.

The defense counsel asked when MD5HASH was performed, the mathematical verification for the hard drive duplication procedure of CaseView. Mr. Elrick said it was done on a routine basis simultaneously with the copying and resulted in a large hexadecimal number. If reverification is done, the new calculation must match the same number to indicate that original material is unchanged. With CaseView and MIRC, he was able to observe the chat logs and configurations.

Defense counsel discussed shutdown procedures for computers with Mr. Elrick. Officer O'Neal had first access to the victim's equipment, but Investigator Willis was the first trained person who viewed the computer. Mr. Elrick confirmed that Officer O'Neal had altered some files on victim's computer. Mr. Vance asked whether pulling the plug was the preferable way to preserve evidence on the crime scene computer. At power-up or shutdown, Microsoft Windows operating system changes numerous files behind the visible screen. Witness Elrick said that normal shutdown may alter files and lose information from memory, so pulling the plug was the better choice between two imperfect alternatives. He agreed that he could not be certain that the copied hard drive was actually the same as the one on the machine at the time of the victim's death.

Ending the in-role activities for the morning conference session, Judge Ford dismissed the mock trial for the keynote plenary session.

Keynote Address: Jurors' Comprehension of Scientific Evidence

Lawrence Solan, Professor, Brooklyn Law School, discussed juror comprehension of scientific evidence used in trials. A study was done to find out how well jurors understood the intended meaning of scientific evidence and, if they did not understand, what other factors affected their decisions in relation to the issues at trial.[15] The study further looked into what might improve jurors' understanding of difficult technical testimony.

During a trial, jurors usually put together mental "stories of what might have happened." A presentation of evidence is expected to fit somehow into such familiar stories. Dr. Solan described how he himself felt when a conference presenter displayed a picture of a hard disk, divided into sectors. He said his attention perked up because he was familiar with that image. He listened more to the narrative. This, he said, is the normal reaction; people will look for concrete instances that are familiar to them and will pay more attention to the person giving evidence at that point.

A Princeton psychologist named Cooper conducted a mock jury trial using a video of "experts' testimony" on a cancer-related issue. Half of the experimental group heard one "expert" who used plain, lay language, and the other half heard the second "expert," who spoke with large medical-technical words, such as "pathological tumor induction." Some of the experimental group were told the expert went to a prestigious university and had published lots of material while the other expert "went to a small State school and was now teaching at a State university." The study found that when the jury understood the language, the credentials of the expert did not matter as much; but if they could not understand, more weight would be given to the expert who had "better credentials."

The study also examined the effect on juror attitude of different levels of payment to the expert. Half of them heard that the expert was being "paid a lot," while the other group was told the expert "was not paid much." Researchers found that, when the expert was not being paid much, then the credential did not matter significantly; but if the juror heard that the expert received a lot of money, they would be more favorable to the expert with the lower credentials. In other words, there was a strong unfavorable reaction among jurors to the idea of an expert as "hired gun." They preferred that the expert be novice at testifying (not frequently in court) and that fees be modest and reasonable.

Professor Solan discussed what had been found to be particularly hard for jurors. Statistical analysis has been shown to be difficult for both judges and juries, although they are "trainable" when steps of the analysis are clearly explained. Counsel has to tell stories that are related to experiences with which the jurors can identify. Particularly hard for lay jurors or the court is the idea of integrating "base rate information" in questions of how probable or reliable a certain event is. There is a tendency, in dealing with jurors, to avoid discussing base rate information altogether.

Open Discussion

Dr. Solan said people often do not understand liability issues in the same way. Judges and juries sometimes accept or reject evidence intuitively despite established analytical procedures in such areas as forensic handwriting analysis. The idea of causation is considered a prerequisite to assigning responsibility. For example, if someone hosts a party and a drunk person leaves, drives a car, and causes a serious

accident, is the party host liable for the damages? If someone leaves a key in the car ignition and a juvenile steals it, then causing an accident, is the car owner liable? Professor Solan said people are evenly divided about how to decide such things. If a person was only an "enabler" for a certain event, some liability, such as a fine, may be assigned but seldom would overall responsibility be assigned.

Responding to a question about the effect of the entertainment industry on juries, he agreed that there was such an influence. For example, some type of "battle" may be portrayed in court cases, such as a rape case in which "lifestyle of the defendant" opposes "criminality of the attacker." He recommended an interesting new book entitled *When Law Goes Pop*.[16]

The group discussed an experiment in jury trials that makes jurors more active, giving them the role of asking the witnesses questions themselves. This appears to increase the quality of jurors' attention. Another participant asked about the possibility of bias introduced by the jury's foreman. Dr. Solan said research by Steve Penman had indicated little influence in terms of changing votes.

In summary, Dr. Solan suggested to the participants that juries could reach a better understanding of technical information if counsel does the following:

- Makes it clear how the evidence fits into the story of what happened.

- Reminds the jury of connections and introduces things they can identify with.

- Trains juries about base rates, if this is important to the evidence.

- Presents technical evidence in tabular form, which is easier for the jurors to compare.

- Remembers, according to other psychological research, that people do better at drawing inferences about what *did* occur rather than about what *did not* occur.

Digital Evidence—Virtual Reality in the Real-World Courtroom [continued]

Scientific, Technical, or Other Specialized Knowledge [In Role]

Judge Judith Ford called the court into session, and counsel for the defense called their expert witness to the stand [played by **Charles Boncelet,** Professor, Department of Electrical and Computer Engineering, University of Delaware]. Dr. Boncelet said he was a professor in electrical and computer engineering and had taught a variety of courses since 1984 in communications, signal processing, networks, and probabilities. He had written a number of articles in his field and had research experience. Dr. Boncelet said he was familiar with features of IRC and the MIRC program, although he had not used the program. It was a prominent application in computer networking.

Defense counsel, Mr. Vance, asked him whether the MIRC logs were text files and therefore easy to alter. Dr. Boncelet said any text editor such as MS Notepad could do so. It would have been possible for a third party to access the victim's computer and alter the log files to redirect suspicion and minimize the chance of being detected in the murder. With MS PowerPoint to display his computer actions, Dr. Boncelet showed participants how to delete or alter the date of entries in a log file using a simple text editor (MS Notepad). He reset the computer's clock to change time stamps assigned to the log by the MIRC software. A third person (possibly the murderer)

might have reset the computer's clock, found and altered log entries, and left misleading "evidence" behind in only a few minutes.

The prosecutor, Mr. Murray, cross-examined, asking Dr. Boncelet if this alteration of the logs could be done by a person who did not know enough to run batch files? Dr. Boncelet said yes, it would only take a little more time in that case. Mr. Murray also explored with him the idea of documents cached by the Windows operating system and then asked Mr. Elrick to return to the stand.

Mr. Elrick described retrieving information from unallocated clusters on the hard drive. When a disk is formatted, all clusters are numbered. A file will be saved into one or more clusters. Each cluster is roughly a page of data, and two files cannot occupy the same cluster at the same time. When the file is deleted, the filename is changed and the space is registered as unallocated, but the content in the clusters is not actually changed.

Defense counsel, in turn, asked Mr. Elrick what material retrieved by CaseView looked like. Mr. Elrick said it would be a text file, but it would also contain information from MIRC about the time the message was sent and who sent it. He agreed with Mr. Vance, however, that it was not necessarily possible to tell who the author of the fragment was.

Summation [In Role]

Mr. Murray told the jury to consider whether necessary standards of reliability had been met. The standards of reliability should be reasonable for the kind of software discussed. CaseView has been designed and tested for law enforcement. It is currently used widely by law enforcement in the field. MIRC, however, has not been tested to the same degree. It is only one of several products of its type, and

special testing of this kind of software has not been considered worthwhile. The market speaks, said Mr. Murray. If law enforcement personnel are consistently buying and using CaseView, this supports the idea that it is a reliable tool. Juries should allow evidence gathered with CaseView, especially since it is not a "weight" standard (deciding guilt or innocence) but just a question of admissibility of evidence. It is preferable to have the greatest universe of evidence available at trial. He noted that prosecution did not object to the demonstration of altering MIRC log files because that, too, had a wider educational purpose.

Mr. Vance, for the defense, asked that the evidence be excluded in its entirety. He said that the Federal Rules of Evidence and case law require a methodology that is relevant and reliable, especially when an individual's liberty is at stake. The computer logs and files pertained to several people, not just the suspect. The government seeks to establish that the defendant communicated with the victim, that there is motive for murder (blackmailing), and that the defendant received information about the victim's whereabouts, giving him an opportunity to commit the crime. However, serious weaknesses have been identified in the procedures yielding this evidence. The product CaseView is shockingly lacking, said Mr. Vance, in terms of independent testing; nor are there any standards, information on error potential, or evidence of peer review of the methodology (except for a magazine article). We know only, he said, that the government uses and likes it. Its reliability is called into doubt when the same people who benefit from it are those who also formally test it.

Another serious weakness in the evidence is the fact that the first officer on the scene of the murder acknowledged making changes on the crime scene computer files. He could have changed configurations or many other things. Investigator

Willis said the safest way to handle an evidence computer of this kind is to pull the plug from the back. But that decision causes the loss of any open files. The prosecution expert, Mr. Elrick, admitted that he could not say with certainty whether the disk he copied and examined was the same as the computer disk at the time of victim's death. A perpetrator with a modest amount of expertise in MIRC software and chat logs could have had the opportunity to access and alter the logs.

Finally, the government's use of CaseView to get portions of text from unallocated disk space fails to meet *Daubert* standards for reliability or methodology. The prosecution cannot know the author, source, or time these fragments were created. Defense asked the court to enter an order precluding entry of all this evidence.

Judge's Instructions [In Role]

Judge Ford instructed the jury to decide whether the prosecution's expert testimony had a sound foundation and should be admitted into evidence. It must be decided whether the evidence is sufficiently reliable to present in a real trial. She said they can take into account factors that bolster or weaken the reliability of the evidence. [*Aside:* Judge Ford also asked juror-participants to comment on approaches of the attorneys, testimony by the proffered experts, and court rulings.]

Judge Ford noted that, since she is from California where *Daubert* is not used, she has no experience with these standards. If she were really on the bench, she would have to "take the matter under submission." This kind of technology increasingly pervades everyday activity, and the judicial system must become ready for it. Counsel in the trial brought forward some interesting considerations, such as the cost and difficulty of standardized testing for software. Judge Ford said that mixed problems, such as authenticity issues and

hearsay, affected this evidence. She asked the following: Should software that is a tool of law enforcement be required to meet different standards from other generally used software? Does software have to pass a higher bar for admission in criminal cases than in civil cases?

Open Discussion

Mr. Vance agreed that many issues were not technically *Daubert* issues, such as the problem of data from unallocated disk space. The defense took a "real world" approach to this, which would have attempted to "take a hit" at prosecution's evidence wherever possible.

One of the participants noted that many of the recent cases involving computers related to child pornography. Most of these have key evidence that consists of saved files, and these would not be affected by questions on reliability of the software. This murder case, however, presents timeline issues, questions about the accuracy of recovery software, and disputes about the context of the evidence. Mr. Murray noted that, when the plug was pulled on the crime scene evidence computer, any material in random access memory (RAM) would have been lost. The best method to document this would have been to do a "dump" of RAM before powering down, but the first responding officer was not qualified to do this. Auto-run programs, such as those intended to defragment the disk, could also have altered what was on the computer's hard drive.

One attendee asked about information that could be received from the ISP. Mr. Murray said ISPs, similar to telephone companies, do not track content. The prosecution did use the ISP to identify Doakes' (the suspect's) computer. The suspect's computer (examined pursuant to a search warrant) had revealed fragments of text (erased files) that contained

user names matching those found in the victim's chat logs. Judge Ford agreed that there were both authentication and *Daubert* problems with the evidence presented in the mock trial.

Jury Findings [In Role]

The jury decided that the evidence was admissible. This decision was reached unanimously. The jurors emphasized, however, that admitting the evidence did not decide guilt or innocence of the defendant. The software used to develop the evidence had been tested by expert witnesses and the police department. There was a general acknowledgment that strict *Daubert* standards could not necessarily apply to software. The jury foreman [played by **Judge André Davis,** U.S. District Court, Baltimore, Maryland] commented that it was not completely clear to the jury whether the evidence satisfied "expert testimony," hearsay, chain-of-custody, or other authentication issues. Also, an inferential step was made by assuming that the defendant had access to the computer and had himself produced the material on the hard drive, although it would typically have been his own computer.

Closing Remarks

Dr. Chaski said this type of case appears in the courts often. She had recently been involved in a forensic linguistics problem relating to a suit against part of the U.S. military. There were disputed documents, and a computer expert testified to show that the documents indeed came from the network claimed by the party. Standard methods of authorship identification can be very hard to apply in the computer field.

Ms. Ballou said that NIST is testing software tools and looking into methods to release reliability testing results without

legal action from the software developers. Corporations often have newer versions that they are trying to quickly improve postrelease. Information from a nonbiased entity would be valuable.

Lisa Forman, Director, Investigative and Forensic Sciences Division, OST, NIJ, noted that scientific technology continues to develop. Less reliable methods must be weeded out for many reasons, but scientific advances can help society find practical remedies to crime. She expressed appreciation to the participants for the interesting display of legal reasoning in these new areas, shown throughout the mock trial, and invited participants to offer suggestions for next year's National Conference on Science and the Law. NIJ will be posting information and proceedings on its Web site and welcomes attendees' responses.

Notes

1. Franklin, B., *Historical Review of Pennsylvania* (1759).

2. *Daubert* v. *Merrell Dow Pharmaceuticals,* 509 U.S. 579 (1993).

3. *U.S.* v. *Van Wyk,* 83 F. Supp. 2d 515 (D.N.J., 2000).

4. *Frye* v. *United States,* 54 App. D.C. 46, 293 F. 1013 (1923).

5. Gatowski, Sophia I., Shirley A. Dobbin, James T. Richardson, Gerald P. Ginsberg, Mara L. Merlino, and Veronica Dahir, "Asking the Gatekeepers: A National Survey of Judges on Judging Expert Opinion in a Post-*Daubert* World," *Law and Human Behavior* 25 (5) (October 2001): 433–458.

6. *Kumho Tire* v. *Carmichael,* 526 U.S. 137 (1999).

7. Hand, Learned, "Historical and Practical Considerations Regarding Expert Testimony," *Harvard Law Review* 15 (1901): 40.

8. Hansen, Mark, "Expertise to Go," *ABA Journal* 86 (February 2000): 44–53.

9. Kovera, Margaret B., Bradley D. McAuliff, "Reasoning About Scientific Evidence Effects of Juror Gender and Evidence Quality on Juror

Decisions in a Hostile Work Environment Case," *Journal of Applied Psychology* 84 (3) (June 1999): 362–375.

10. Loftus, Elizabeth, and James M. Doyle, *Eyewitness Testimony, Civil and Criminal, Third Edition,* Charlottesville, VA: Lexis Law Publishing, 1997.

11. Wigmore, John H., "Professor Munsterberg and the Psychology of Testimony," *Illinois Law Review* 3 (1909): 399–445.

12. Risinger, D. Michael, "Defining the Task at Hand: Non-Science Forensic Science After *Kumho Tire Co. v. Carmichael,*" *Washington and Lee Law Review* 57 (3) (Summer 2000): 767–800.

13. Ladd, C., M.S. Adamowicz, M.T. Bourke, C.A. Scherzinger, and H.C. Lee, "A Systematic Analysis of Secondary DNA Transfer," *Journal of Forensic Sciences* 44 (6) (1999): 1270–1272.

14. A "jury" of 12 audience members was selected at random from volunteers who submitted business cards for a drawing. This would not be a typical jury because it consisted of three judges, several crime laboratory directors, Ph.D.-level researchers, and criminal attorneys.

15. Vidmar, Neil, and Shari Seidman Diamond, "Juries and Expert Evidence," *Brooklyn Law Review* 66 (4) (2001): 1121–1180.

16. Sherwin, Richard K., *When Law Goes Pop: The Vanishing Line Between Law and Popular Culture,* Chicago: University of Chicago Press, 2000.

Fourth Annual Conference on Science and the Law, October 3–5, 2002

Thursday, October 3

Preconference Workshop I: Interpretation of Scientific Analytical Reports

Elements and Interpretation of an Analytical Chemistry Report

José R. Almirall, Director, Forensic Science Graduate Program, Assistant Professor, Department of Chemistry, Florida International University, noted that responsibility for correct interpretation of scientific analytical reports lies both on the writer, who must write an understandable report, and on the reader, who must do the background work to understand the language. Language is the crucial element in any disconnection between the writer and reader. Scientists and attorneys alike have their own languages with specific terminologies. Additionally, reports are not always clear, and the reader may not be prepared to understand the material. The clarity and completeness of reports is important: Were conclusions reached? Do examinations and results warrant the conclusions? Does the scientific community agree?

A good report will include identification marks (agency, date, examiner who conducted the analysis, etc.), a description of the items examined, a description of the examinations conducted, and an interpretation of the results. For example, in analytical chemistry, a measurement science, the correct methodology/tools for qualitative and quantitative analysis must be shown. In choosing samples for analysis, which requires determining what to analyze and how much to analyze, procedures must support the interpretation of data and offer feedback for decisions.

Standard terms in scientific reports include the following:

- **Precision.** How closely measurements of the same quantity come to each other.

- **Accuracy.** How close experimental observations are to the true value.

- **Uncertainty.** Reporting of numerical values (including random error value).

- **Bias.** Systematic errors related to accuracy.

- **Detection limits.** Smallest quantity that can be detected and reported.

- **Quantification limit.** Smallest amount that can be quantitatively measured.

Dr. Almirall showed two examples of controlled substance analysis reporting, with correct examination procedures, interpretation of data, and reports discussed in the guidelines and standards of the American Society of Crime Laboratory Directors (ASCLD) and ISO 17025 as well as the recommendations of the Scientific Working Group for the Analysis of Seized Drugs (*www.swgdrug.org*). He indicated that controlled substance analysis should include qualitative identification of the compound, purity, weight, uncertainty of measurements, all methodology, and

41

resulting data. He also described reporting on other trace evidence analysis and the use of the Fire Debris Evidence Guidelines of the American Society for Testing and Materials in seeking to identify accelerants in complex mixtures. When seeking to identify accelerants like gasoline, the report should show extraction samples from the matrix, analytes, and peaks that could be expressed. Interpretation should consider where the samples were found.

The Scientific Working Group on Materials Analysis (SWGMAT) works on cases involving other trace evidence, such as glass transfer (e.g., when a window is broken during the commission of a crime, the forensic examiner may look for glass in clothing that may match the window onsite). In a two-tiered approach, analytical chemistry first looks for a "match" for the sample, then calculates how common or uncommon the material would be in the location where the sample was found.

Scientists interpret evidence using their experience, knowledge, statistical research, and literary documentation. A lawyer should check to make certain that a laboratory has received ASCLD accreditation and follows accepted standards and procedures. A "match" found through analytical chemistry means something different from the term "match" that is used when comparing evidence in court. In an evidence report, descriptive words such as "consistency" and "distinguishable" are preferable to "match." Usually, forensic scientists are given the flexibility to choose among methods, and the report for court is a summary rather than a compilation of all procedures.

Interpretation of DNA Reports

Jack Ballantyne, Associate Professor of Chemistry, University of Central Florida, said that laboratories preparing DNA and serology reports need a documentation toolkit containing the following publications and guidelines:

- DNA Advisory Board, *Quality Assurance Standards for Forensic DNA Testing Laboratories.* (*http://www.fbi.gov/ congress/congress02/forensicstd.htm*)

- Scientific Working Group on DNA Analysis Methods (SWGDAM), *Short Tandem Repeat (STR) Interpretation Guidelines.* (*http://www.fbi.gov/hq/ lab/fsc/backissu/july2000/strig.htm*)

- National Research Council II, *Evaluation of Forensic DNA Evidence: Update on Evaluating DNA Evidence,* National Academies Press, 1996. (*http://www. nap.edu/books/0309053951/html*)

- John M. Butler, *Forensic DNA Typing: Biology and Technology Behind STR Markers,* San Diego: Academic Press, 2001. (Not available online.)

In addition, the toolkit should include specific, maintained laboratory procedure manuals showing who conducted each analysis and guidelines for interpretation and report writing. A wide variety of report formats may comply, but he recommended routinely enumerating the significance of test results (attaching a number) and using additional interpretation elements that have been standardized and described by SWGDAM.

DNA short tandem repeat (STR) typing is done with fluorescent STR kits, and separations are prepared with capillary electrophoresis. Genotyping is done by comparison of each locus, which has two alleles (inherited from mother and father). If the example is heterozygous, the alleles will be different sizes; if it is homozygous, they will be same size. The test measures relative fluorescence unit (RFU) values indicating the alleles and allelic ladder. A sample comes from a single source if each typed locus has only one or two

alleles. The determination of exclusion or inclusion is based on all tests and observations made for that sample. When the analyst works with degraded results or starts with low quantities of DNA, observations are more open to question.

Existing population databases list allele frequencies within the general population (Caucasian, Hispanic, African American). Statistics for known frequencies in population genetics can indicate genetic variation in general populations, frequency of homozygous or heterozygous DNA, and the occurrence of a particular genotype in the setting where a sample was found. Laboratory analysis conclusions about a questioned sample possibly containing the same genotype and a known sample are expressed to a reasonable degree of scientific certainty. Each laboratory sets its own keratin thresholds for empirical analyses, such as RFU of 150. The term "match" is used to qualify the comparison of samples (allelic comparison). The results obtained from a questioned sample are compared with the results from the known samples in the case; the determination of inclusions or exclusion is the responsibility of the analyst working the case. The analyst is required to list the loci that were tested. Exclusions are more clear than inclusions or inconclusive results. They will show a difference in at least one locus, although results may be inconclusive if threshold standards are not met at one or several loci.

Presumptive tests determine the presence of blood, semen, saliva, etc., and definitive tests (blood type, presence of spermatozoa, etc.) may follow. If a mixture of DNA is present, there can be three, four, or more alleles at each locus rather than two. If a dominant profile can be determined in a mixed pattern, the laboratory report will state the major and minor contributors. Otherwise, the analysis calculates and reports frequency determinations for possible combinations.

Dr. Ballantyne noted that the location and composition of the biological material can be as important as the DNA results themselves.

Forensic Document Examiners Report: What Lies Within

Gerald B. Richards, owner, Richards Forensic Services, said that handwriting examination is pattern recognition and relies on good comparisons. Often the forensic scientist, lawyer, juror, and judge understand certain terms differently. The Scientific Working Group for Forensic Document Examination (SWGDOC) is working on the following nine-level standard probability-related terminology for conclusions relating to questioned documents:

- Identification.

- Strong Probability.

- Probable.

- Indications.

- No Conclusions.

- Indications Did Not.

- Probably Did Not.

- Strong Probability Did Not.

- Elimination.

Report formats include date and case information, a description of materials submitted (questioned, known, and specimen materials), a list of the questions being asked (what examinations are requested), the results of the examinations (the answers offered), and disposition of the submitted materials with any additional remarks.

Mr. Richards gave demonstrations of good and bad document examination reports. The date on which the report was written,

the date of communication to the examiner, and the date on which the materials were received should be within a reasonable range of each other. Materials must be described so that each item can be individualized and identified by itself. Questions discussed in the report should be pertinent to the case. If a computer program is used as a primary document examination procedure, it should form part of the report. Technical jargon should be kept to a minimum, except to explain procedure. The results of the examination should clearly and concisely state what has been accomplished and provide additional information that might be valuable.

Principles of Forensic Mental Health Assessment

Kirk Heilbrun, Chair, Department of Psychology, Drexel University, noted that forensic mental health assessment, which is performed to assist the court in legal decisionmaking or is presented as evidence in a legal case, contrasts with therapeutic assessment, which is performed to assess a patient's symptoms and guide the planning of mental health interventions on the patient's behalf. The assessment report should first identify relevant forensic issues and legal questions to be cited. The mental health examiner should accept referrals only within his or her area of expertise and decline any referrals in which the examiner's impartiality might be questioned. The examiner's professional degree, licensure, board certification, curriculum vitae, and summary of qualifications must be available if requested.

In the report, the role of the mental health evaluator should appear clearly as court appointed, attorney requested, or other consultant. Financial arrangements for the support of the evaluation should be clarified. A combination relationship (a clinician who serves as both therapist and forensic evaluator) is particularly undesirable. Such

a blend would be proper only with explicit justification, advance planning, and clear notification to the individual being evaluated.

Relevant historical and clinical guidelines and data sources support the assessment of clinical characteristics in reliable and valid ways. Using accepted elements of notification and consent, the expert gives appropriate notification of the purpose for the examination and obtains authorization before beginning.

Forensic mental health assessment often follows certain models, such as the Morse Model, to guide data gathering, interpretation, and communication. The assessor determines whether the individual understands the purpose of the evaluation and explains the associated limits of confidentiality. He or she has to ensure that the conditions for the evaluation are quiet, private, and distraction-free, noting any deviation and describing any outside impact on the data collected. The report uses valid data sources and places relevant historical information in a separate section to document the individual's history and previous functioning in areas relevant to the current clinical condition. Third-party information can help in assessing response style when compared for consistency with self-reported information. The examiner has to assess legally relevant behavior and document the individual's functional legal capabilities.

The report shows scientific reasoning to assess the individual clinical condition, covering functional abilities and legal/causal connections. Dr. Heilbrun cautioned that, during testimony, the mental health evaluator should describe the findings and their limits in such a way that the evaluator need not change testimony during cross-examination. The assessment presents conclusions about forensic capacities but does not address the larger legal questions. In testifying, the assessor bases his or her testimony on results of the selected

instruments and relies on the contents of the report to guide that testimony, using plain language for lay listeners in court.

Preconference Workshop II: *Brady* and Other Ethical Issues Facing Forensic Scientists

The case of *Brady* v. *Maryland*[1] points to a duty on the part of prosecutors (and their "law enforcement team") to turn over evidence that is material to the defense. Moderator **Barry A.J. Fisher,** Director, Crime Laboratory, Los Angeles County Sheriff's Department, noted that several questions are raised: What is material exculpatory evidence? Whose duty is it to find it? What is the crime laboratory's responsibility?

Watershed cases involving laboratory misconduct (*Zain*[2] and the professional sanctions against Joyce Gilchrist[3]) have caused people to ask what is the correct response to scientific misconduct. Responses have included employee discharges, civil litigation, and even attempts to prosecute for criminal perjury. Thus far, courts have interpreted the variety of questions raised by these cases conservatively, ruling that evidence that was allegedly tainted by laboratory misconduct must be "material" to the specific case in question. Prosecutors need to understand laboratory work better, and defense attorneys need more education in scientific areas and more open attitudes toward discussion. In general, the *Brady* case has been an influence for better cooperation between adversaries in the process of forensic discovery.

Paul C. Giannelli, Weatherhead Professor of Law, Case Western Reserve University, said *Brady* is sometimes seen as asking the prosecution to "hand over evidence and assist the accused." Where the prosecutor really believes the accused is innocent, the case is likely to be soon

dropped. In practice, the *Brady* responsibility on the part of the prosecution refers to evidence that would be "almost outcome determinative." The decision has resulted in more discovery by defense attorneys under the Freedom of Information Act and more efforts to find out about misleading laboratory information.

Professor Giannelli described many experiences of courtroom error, ranging from failures to use available DNA samples to forged fingerprints. He divided the relevant issues according to courtroom role (prosecutor, expert, defense counsel) and type of problem (misleading credentials, missing evidence, misleading statistics or reports, late disclosure, pressured opinions, and exculpatory evidence in the prosecution's possession). He said the decisions mainly targeted incompetent defense counsel, prosecutors who have not taken the "assume innocent" duty seriously, and laboratories that have a propolice orientation.

The laboratory report is a scientific document and must clearly specify what tests were used, what results were observed, why the particular methods were chosen, and whether the results have limitations. Discovery of laboratory notes and graphs ordinarily would be expected. In a large jury case, laboratory reports and protocols have to be understandable to laypersons and would best be posted to provide both sides with easy access. A system should be designed so that an average prosecutor and average defense attorney with typical caseloads could use the material productively (an example would be making certain key documents available for examination via the Internet).

Albert J. Krieger, Chair, Criminal Justice Section, American Bar Association, said the defense attorney typically has no specific scientific understanding and is confronted (sometimes immediately before trial) with prosecution's expert scientific

witness. Mr. Krieger said, "You might call this 'trial by ambush.'" The defense counsel needs an adequate breadth of discovery for scientific evidence, and it is not unreasonable to require the prosecution to share materially important evidence. Sometimes comparative studies on scientific materials say contradictory things about the same evidence. State money is well spent in such circumstances when it gives defense attorneys relief under *Brady*.

A defense attorney's gut reaction to charges against the defendant will be, "You have to prove it." He or she has the right to explore any open questions in the interest of justice. The justice system requires candor and honesty from witnesses to arrive at court decisions that give closure in complex cases. Any kind of deliberate misstatement in court is inexcusable and criminal. The idea that a scientist would give testimony to "serve an agency" is an affront to the system.

Haskell M. Pitluck, Retired Circuit Court Judge, Illinois 19th Circuit, reviewed codes of ethics in the context of forensic casework. Judge Pitluck, past president of the American Academy of Forensic Sciences (AAFS) and member of the Council of Scientific Society Presidents, said most such societies (similar to the AAFS) already have some kind of code requiring that members refrain from behavior adverse to the society, that they do not misrepresent any criteria for membership in the society, that they do not misrepresent study data, and that they obtain permission from the society before using its statements. Violating such guidelines is usually grounds for censure and expulsion.

Should lawyers check the credentials of a potential expert witness? Yes: Most ethics cases are brought because an expert's credentials have been misrepresented. Many professional organizations have

procedures for hearings on violations. However, incompetence is not necessarily the same as unethical behavior.

Rockne P. Harmon, Senior Deputy District Attorney, Alameda County, California, told the group that to say that prosecution decisions are "not reviewable" is inaccurate; prosecutors want to reach good decisions that will not be reversed. Photo lineup results, laboratory reports, witness statements, codefendants' statements, and even evidence regarding weather may or may not be considered "material" to the defense counsel. The Maryland Supreme Court has affirmed the duty for disclosure under *Brady* only for a death penalty case. The keywords are "favorable" [to the defendant] and "material" [to the case]. The debate must be understandable, about case-specific critical evidence, not report-writing standards, etc.

Forensic evidence discovery for cases is designed to ensure a fair trial according to the rules of law. The requirements under *Brady* have resulted in "game playing" among defense counsel, sometimes leading to large additional areas of discovery for secondary information that is circumstantial, such as high school or college transcripts of witnesses, complete proficiency testing details of a laboratory, or lists of every other time a witness has testified.

The requests for reanalysis of tests or for credentials and accreditation information must have meaning in the specific case context and must be material to the guilt or punishment of the defendant. Many times, the prosecution (not only the defense counsel) will suggest retesting DNA evidence from the defendant, and defense may oppose this. *Brady* is a workable restriction, but it has been influenced too often by consideration of gray areas outside the scope of a specific case.

Preconference Workshop III: Can DNA Be the Magic Bullet? What DNA Can (and Cannot) Do

The Good, the Bad, and the Ugly: Realities of the Forensic DNA Laboratory

Cecilia A. Crouse, Serology Section, Palm Beach County Sheriff's Office, said that the good news is that crime laboratories have developed good, solid accreditation standards and many organizations (including the National Institute for Standards and Technology [NIST], National Research Council, and American Society of Crime Laboratory Directors) are now involved in ensuring quality forensic DNA testing. The bad news is that the laboratories are unable to handle all documentation and work every case. Interpreting and reporting DNA profiles requires specific qualifications. Official DNA guidelines now address planning and organization of the laboratories, personnel, documentation/manuals, equipment/facilities, evidence-handling procedures, analytical procedures, and casework documentation. Peer-reviewed professional journals (*Journal of Forensic Sciences, Forensic Science International*, and *International Journal of Legal Medicine*) are established. New standards require procedures for qualifying personnel, maintenance of technicians' qualifications, and special requirements for supervisor/technical leader positions, examiner/analysts, and technicians. DNA manuals covering laboratory documentation rules and recommendations for validation, procedures, equipment handling, and technician qualifications have developed substantially since the end of the 1980s.

Information on base populations has grown with the STR database and the ongoing work of the Scientific Working Group on DNA Analysis Methods, aided by FBI national standards and uniform audit documents. Remaining analytical problems involve microvariances, low copy number (small amounts of genomic DNA), and challenges in mixed samples. In mixture cases, the analyst must ask, "Could this mixture include both parties based on the witnessed allele peaks?" Examination must identify the number of potential contributors, estimate the relative ratio of individuals contributing to the mixture, consider all potential genotype combinations, and compare to reference samples.

Computer-assisted DNA data interpretation can offer caseworking laboratories a way to reduce backlogs. However, the sources of biological evidence and the sheer amount of evidence collected have become overwhelming for many laboratories. In a violent crime case, there may be about 5 submissions for analysis, 10 items screened, and 20 stains examined. STR kits have also changed as technology has developed. They now require validation testing, and up to several days may be needed for secondary review. Laboratories need expert software simply to manage the large case-processing backlog.

DNA as a Magic Bullet

Mark E. Windham, Head Deputy Public Defender, Los Angeles County Public Defender's Office, and **George "Woody" Clarke,** Deputy District Attorney, San Diego, discussed the idea that DNA is a "magic bullet": The odds for an STR DNA match are 1 in 1.78 quintillion people, for example, which can create an impression of overwhelming evidence against a defendant. Defense attorneys sometimes do not ask for independent DNA testing because faith in this evidence combines with occasional DNA test results that contain serious errors, possibly due to microvariants, mixture analysis issues, artifacts or "stutter" in the analysis, contamination, or employee misbehavior (as in the

Gilchrist and *Zain* cases). The common assumption that "DNA [evidence] exists so the case is over" might have dangerous results.

New standards for evidence began to be applied after the *Daubert*[4] decision, which required a preliminary assessment of evidence for admissibility. However, courts still have not decided about the best treatment of mixed-DNA evidence. Not all states have used the *Frye*[5] standard for evidence (1923 rule on general acceptability of scientific evidence). In the Danielle van Dam case,[6] DNA evidence removed lingering doubt among jurors; about 10 percent of jurors indicated they would use DNA evidence to decide to impose the death penalty. Some States require "likelihood ratios" or statistics to be introduced in cases for which DNA evidence from a mixed specimen has been admitted. Statistics are required in any DNA case.

Forensic analyst qualifications form another source of contention. Witnesses in court are not necessarily the bench analysts who have performed the tests. Single-source DNA is the most powerful and trusted evidence for identification, but increased sample collection has caused more use of mixed DNA samples. STR-based systems, used to separate mixtures, are perhaps less questioned since they are similar to the older polymerase chain reaction (PCR) testing that first successfully gained court admissibility. Juries are becoming more confident in evaluating testimony that includes frequency data, discussions of error rates, and consideration of the probable reliability of specific scientific tests. Most forensic laboratories keep statistics on their test findings and use agreed-upon standards.

Welcome and Opening Remarks

Sarah V. Hart, Director, National Institute of Justice, U.S. Department of Justice, welcomed attendees to the Fourth Annual Conference on Science and the Law, noting that NIJ's mission of advancing science, technology, and program evaluation was primarily intended to serve State and local governments rather than the Federal Government. The capacities developed in criminal justice systems through the science of DNA have stimulated great debate not only in the laboratories but also among police, courts, trainers, law enforcement technical assistance personnel, and taxpayers.

Director Hart said that information on the public benefits reaped from use of DNA evidence has not been well captured. From the perspective of a police chief's investment, she asked, is the right information being made available? How much safer is the public because of the use of DNA evidence? Recently, the U.S. Attorney General committed significant funding for DNA research; this is evidence of the administration's commitment, but research needs to determine whether technologies are being used to their maximum potential. Are the hard questions (e.g., those currently concerning use of fingerprints and other trace evidence) being addressed clearly, in a way that will benefit the courts and justice system?

Keynote Address: DNA and Genetics

Ming W. Chin, Associate Judge, California Supreme Court, San Francisco, noted that science and the law use different languages and have different objectives. In

science, an error and a mistake are not the same thing. Scientists must include analyses of errors in their writeups of an experiment, but anyone may make mistakes and then simply clean up and try again. In law, an error ("mistake") may mean overturning a decision, with large implications. Science is aiming at truth; the law is aiming at justice. Science can allow perpetual revision because it is not subject to time limits, while the legal world prefers a decision to be final and quick. However, much intersection and conversation occur between the fields. Both disciplines seek to arrive at rational decisions and to avoid the influence of self-interest.

DNA evidence already plays an important role in criminal cases. Approximately 100 death row inmates have been exonerated on the basis of DNA testing. More controversial is the discussion of how, why, when, and for whom the DNA databases need to be maintained. States have approved these databases, but they are not uniform: Some States require DNA testing at the point of arrest, some on indictment, some for felons only, and some for sex offenses only.

Genetics may be expected to have more influence on the law as research continues. Scientists have now developed about 900 screening tests for genetically caused diseases. Great benefits and great problems must be considered in this scenario. Privacy loss and classification of people according to their DNA could lead to grave ethical risks. Fair employment practice, reproductive technology, and life itself may be challenged by these changes.

Among Fortune 500 companies, 84 have admitted to using medical records to make hiring, firing, and promotional decisions. Insurance companies already use information for coverage decisions from the 900 available genetic tests, if they are able to find such results. On reproductive issues, California has declared that stem cell research offers great promise; in Washington, discussion has gone in exactly the opposite direction.

We must have open minds to meet challenge and change, noted Judge Chin, so that conversations about genetics offer more understanding to justice system participants as they try to "catch up with the science." We will be dealing with these conflicts for years. History teaches that scientific progress is inevitable and unrelenting. The more we prepare each other and discuss these issues, he noted, the more able we will be to find reasonable solutions.

Annual Reports on Science and the Law

Moderator **Chief Justice Shirley S. Abrahamson,** Wisconsin Supreme Court, noted that the theme of this conference is communication. Scientists, lawyers, judges, and other nonscientists must be able to understand each other. Many decisions made by legislators, administrators, and judges affect science and are influenced by science in turn. Laypeople must be able to make decisions from an educated standpoint.

Post-*Daubert*: Is the Sky Really Falling?

Donald Kennedy, editor-in-chief, *Science* magazine, American Association for the Advancement of Science, said there are many forms of science in addition to *Daubert's* hypothetical view. Dr. Kennedy discussed the boundaries of what most people call science and the legal system's treatment of clinical experience and scientific research in testimony. The use of DNA evidence in forensics has cast a strict scientific context on other kinds of trace evidence and extended the territory across which lawyers and scientists must communicate effectively.

The nature of science fosters rapid change, but law works on precedent, requires more time, and has to try to resolve disparities. The Supreme Court has supported a number of evidentiary standards, including *Frye*, *Daubert*, and peer review. Dr. Kennedy urged that the courts refrain from placing too strong an emphasis on peer review. Also, he did not consider court-appointed experts to be good substitutes for experts brought by the parties (with full notifications and under judges' discretion).

Regarding litigation-sponsored science, Dr. Kennedy said that the current confusion will probably give way to a greater quantity of needed research that will lead to a disposition of issues that now create conflict. It would be unwise, he thought, to discourage incentives to use science. "We ought to have more research, not less," he noted, "and such research should not be disqualified merely because it is proffered in litigation, if it passes other tests."

Recent Issues in the Courts

David Faigman, Professor of Law, Hastings College of Law, University of California, gave an overview of recent issues that have created something like a "scientific revolution" in the courts. Procedural and substantive questions have been related to changes and increased rigor in technical fields that affect the testimony of police officers, clinical medical testimony, and other testimony. Courts have been experiencing changed powers of review and more use of court-appointed experts.

The *Kumho Tire*[7] case brought attention to the rigor in a variety of technical fields and their methodologies. For example, handwriting evidence and fingerprints have been receiving increasing challenges in the courts, and such areas as firearms identification methods and police field sobriety tests have been examined more closely. Engineering procedures have been rejected especially often with requests for explanations as to why alternative causes should be excluded.

How should submissions be tested? What qualifications are needed in which context? Some theories, in psychology for example, are accepted therapeutically, but not in the court for forensic purposes. Typically, police or Federal law enforcement personnel are not asked to explain the basis of their experience when they testify about illegal drug trade or organized crime, but a handwriting expert's testimony may be accepted in one jurisdiction and thrown out of a court in a nearby jurisdiction. More and more, Professor Faigman said, the appellate courts are being asked to "smooth out differences" and resolve such inconsistencies.

Open Discussion

During open discussion, the group considered statistical methods and how standard samples are created. Scientific validity and validity in a legal argument mean different things. Judge Abrahamson noted that some observers say scientists dislike cross-examination, and courts and juries are unable to deal with scientific evidence; therefore, special courts should be arranged. She asked the audience how many agreed with this view; a large majority of those attending the session favored the current adversarial system. Professor Faigman added that, in his view, science works better in the courtroom with more independent neutral experts and greater judicial control. The decision to admit evidence is a policy choice and has to be seen in the context of what practically may be tested and what is at stake (life or liberty?). "*Daubert*," he said, "if it did nothing else, put the fear of God (gatekeeping judges) into the professional communities."

Scientific and Legal Tutorials (formerly "Science 101" and "Law 101")

Judge Ron S. Reinstein, Superior Court of Arizona, moderated this group of presentations. He said that many are uncomfortable with "unholy" partnerships among scientists, lawyers, and judges. Often people do not use helpful communication skills that are readily available. "Preparation is so important," Judge Reinstein said. "Use of an independent scientist to conduct a short tutorial can assist in a complex case."

Expert Evidence: Basic Rules and Contemporary Controversies

Jennifer Mnookin, Associate Professor, University of Virginia Law School, spoke about the meaning of *Daubert* and *Kumho* in relation to fingerprinting, handwriting, and trace evidence analysis. Trial judges have been given a greater obligation to see that scientific evidence is relevant and reliable. Evidentiary reliability is connected to concepts of "valid science."

The *Daubert* criteria have been used almost as a definitive checklist: testability or falsifiability, existence of standards, known or potential error rates, and relevant peer review or publications. Under *Frye*, courts ask, is there general acceptance in the relevant technical community? The cases have invited courts to be less deferential to scientists solely on the basis of credentials and have pointed out the fairly large discretion judges could use to look at the science itself and to consider its validity. Since these cases, use of pretrial hearings and hurdles for admissibility in some areas have increased. The decisions have also brought more questions of causation into the civil trial context and reduced "junk science," allowing courts to throw out weak cases at the summary judgment stage. The disadvantage for

judges and other participants in the judicial process has been a lack of consistency among different courts and jurisdictions. Throwing out the right cases can protect the system from long expensive trials and troubling outcomes, but conflicts of experts have been a broad concern.

Additionally, many forensic sciences have not emerged from the university-based research tradition where controlled experiments were performed. Epistemological issues have arisen: What should count in the court as expert knowledge? Questions increasingly arise in relation to handwriting analysis, psychiatric evaluations, clinical medical tests, and eyewitness evidence. In the past year, about 40 court challenges have been raised with regard to fingerprinting evidence. Some courts have even begun to exclude handwriting evidence automatically.

Investigating, Evaluating, and Preparing Experts

Carol Henderson, Professor of Law, Nova Southeastern University, spoke about the importance of avoiding the "inexpert witness." A preliminary question must be considered: How do you know that the witness is who he says he is? Recently, the *American Bar Association Journal* has covered quite a few cases of misbehavior by expert witnesses, arousing increased scrutiny and accusations of fraud, negligence, or even criminal malfeasance. Of the 4,600 hair and fiber cases that Oklahoma City police chemist Joyce Gilchrist worked on, 23 were death penalty cases, which stimulated strong public criticism. Problems with forensic evidence have also occurred in other countries. In the United Kingdom, for example, a government inquiry has usually been held after the problem is found to try to fix it and neutralize its effects.

Judges are becoming more mistrusting toward both the "expert witness" and

"peer review" as standards, noted Professor Henderson. Many more hearings are being held on expert witness evidence and such evidence is being excluded more frequently. Surveys of jurors confirm a similar trend among juries. In a 1998 survey, 50 percent of jurors believed that expert witnesses say only what they are paid to say. More disturbing, 75 percent said they would cast aside a judge's instructions about the law concerning expert witness testimony to hand in a verdict that they felt was right. Some kinds of evidence, notably probability evidence in DNA matters, are given less weight than they received previously due to jurors' concerns regarding laboratory errors.

Professor Henderson described the perfect expert witness as one with good credentials, no apparent bias, a pleasant personality, a presentation that is not overly long or complicated, and effective visual aids. American cultural aspects (pertaining to "Generation X" for example) have affected the courtroom. Particular obstacles to juror understanding include use of jargon, lack of visual aids, and juror preconceptions based on TV influences (such as "CSI Miami"), which give jurors the idea that they already know how forensic examination should be conducted. Recent studies indicate that the average attention span of Americans is 1.5 minutes. To be effective, expert witnesses (and attorneys and judges) must make a quick, effective impression on the jurors. Surveys have demonstrated that jurors evaluate witness credibility based on verbal skills, vocal skills, and appearance and body language. The latter carry the greatest weight with jurors.

Professor Henderson discussed the concept of "expert shopping," saying that it makes sense to check available databases, some of which are online, to investigate experts' backgrounds. In one recent study, one in three people had falsified credentials or at least altered them in some way. Too many organizations now offer diplomas for sale or "checkbook credentials" that say nothing about real skills. She also recommended that attorneys check available transcript database resources to review the expert's previous testimony. In Canada, Australia, and the United Kingdom, royal commissions also have begun to study problems relating to questioned evidence.

A recent Federal Judicial Center study indicated that judges are more effective gatekeepers in civil cases than in criminal cases. Attorneys have to prepare an expert for trial and meet with the expert. As she recommended pretrial meetings with an expert, Professor Henderson noted that expert witnesses often say they were called at the last minute and that the lawyer never met with them or reviewed their report until they were walking to the courtroom. If time allows, attorneys should conduct a mock cross-examination and bring up points likely to be raised. Tools to find the needed expertise and good preparation are the keys to successful expert testimony. Professor Henderson described a new online resource under development that pulls together law, science, and technology information for such areas as fingerprints and digital image enhancement.

Toxic Torts

Bernard Goldstein, Dean, Graduate School of Public Health, University of Pittsburgh, examined discomfort among scientists involved in the legal process. The legal approach is one-sided and adversarial by nature, while scientists experience ethical doubts about supporting only one side of an issue. Scientists also dislike being alone in their opinions and look to other scientists for consensus and credibility. He said that attorneys should be sure that scientists understand the rules under which they are "playing" in the courtroom.

For example, the "truth" is only what has been specifically requested and what the judge will permit in the courtroom. Political and legal systems have difficulty with ambiguous, uncertain, or complex circumstances. Some key areas in defining an expert witness include educational background, work experience, publication record, professional memberships or fellowships, board certification, and service on national and international advisory committees.

Causality in toxic tort cases may be generic (e.g., Can this chemical cause the effect?), or specific (e.g., Was this exposure more likely than not to cause the effect?). Dr. Goldstein said "dose makes the poison," among humans as with other animals. Chemicals have specific effects in specific quantities, and scientists legitimately differ on how they rate the effects. Testifying in court, they take the weight of evidence and base their decision on a continuum of evidence.

Criteria for judgment about health risks from possible exposure to toxic substances include strength of exposure, consistency, specificity, temporality, dose response, and biological plausibility. Risk assessment is increasingly used in government, regulatory organizations, and toxic tort cases. In toxic tort lawsuits, for example, claims may try to extrapolate from a dose that is known to cause leukemia to a dose to which the plaintiff knows he or she was exposed. Extrapolation techniques, time-dependent relation-

ships, and possible cumulative risks are considered because society asks to be protected against low-level health risks. Cancer-causing chemicals (which might be threatening only in a high dosage) are treated differently than dangerous materials that accumulate and stay in the body, such as mercury.

Juries have trouble understanding that statistical laboratory norms differ from clinical norms. To "get into court" with a post-*Daubert* toxic tort case, support must exist for an inference of causation. In some courts, this requires a relative risk of 2.0 or greater in an epidemiology study. That is, it must be at least "as likely as not" that the cause of injury was what the plaintiff alleges. Shortcomings for establishing an odds ratio in occupational epidemiology cases include the healthy worker effect, inadequate exposure data, dilution of a high-risk group, and appropriateness of the time period. The "golden criterion" for scientific acceptability is replicability.

Closing Remarks

Lisa Forman, Director, Investigative and Forensic Sciences Division, Office of Science and Technology, National Institute of Justice, spoke about practitioners' search, in both law and science, for facts and assumptions. The law wants "truth without ambiguity," while science practitioners often see ambiguity (strengths and limitations of data) as truth itself. Appropriate control groups in tests, use of commercially prepared DNA as "molecular rulers," and detailed protocols can validate methodology to the court.

Friday, October 4

Counterterrorism: Plenary Panel

Parney Albright, Senior Director for Research and Development, Office of Homeland Security, Washington, D.C., and **Shana Dale,** Chief of Staff and General Counsel, Office of Science and Technology Policy, Executive Office of the President, discussed how science fits into organizational efforts for homeland security and the effects these policies have for the scientific community. In the United States, science and technology skills represent an area of asymmetric advantage. The Office of Homeland Security is seeking coordination of companies, institutes, universities, government labs, and more to consolidate fragmented efforts for greater security. "Enhance the commonplace," using a system perspective, will form a theme. Dr. Albright's agency recognizes that it is difficult to sustain security efforts across the country if these efforts do not also enhance people's everyday activities.

According to the proposal before Congress, U.S. Department of Energy national laboratories, other Federal laboratories, universities, and the private sector will become part of a national research and development enterprise focused on homeland security issues that will be organized around portfolios relating to many kinds of research overseen by the Office of Homeland Security. People will interact on key research topics in chemistry, physics, life sciences, engineering, environmental science, social/behavioral sciences, space, and telecommunications. Scientists will pursue enduring core research and development activities and maintain relationships to port authorities, borders, and security agencies important to the Nation.

Biometrically measured travel documents are planned as part of a new entry-exit initiative affecting foreign travelers. Biometrics will be taken as part of the visa application process, and applicant information will be checked against various "watch list" databases. The legitimate flow of traffic through borders and tourism will be affected as little as possible. A timely, quick way to check people at borders is planned.

Ms. Dale spoke about heightened awareness of vulnerability to terrorists. Privacy rights have to be balanced against efforts to protect the Nation. In cases of students and visiting scholars, when sensitive knowledge could threaten the United States, the administration is implementing enhanced case-by-case review. Research efforts in human health and the agricultural sector (such as virus sequencing for animals and people) will be converging. The National Institutes of Health (NIH) and the Centers for Disease Control and Prevention (CDC) are developing new diagnostics and vaccines relating to human, animal, and plant pathogens. "Select agent" registration for about 60 substances or agents is in development, with corresponding laboratory security for dangerous microbial and biological materials and sequenced reference strains for each required

microbe. Appropriate available information for research and education as well as enforcement of proper use and transport methods for the lists of fungi, viruses, etc., make up part of this program.

Considerable cooperation of State and local authorities is needed, noted Dr. Albright. Institutions and information structures must be designed to coordinate at every level. The fight against terrorist attacks actually takes place among State and local responders, who will be responding with new equipment and methods. Audience discussion introduced the need for greater support for State crime laboratories and their training.

When Is Evidence Considered Manipulated? A Close Look at Digital Evidence: Plenary Panel

Moderator **Susan M. Ballou,** Program Manager, Forensic Science Projects, Office of Law Enforcement Standards, National Institute of Standards and Technology, noted there has been a growing discipline connected to digital records and procedures for determining when images have been "tainted" for evidence purposes.

John A. Boesman, Detective, Computer Forensic Unit, Prince George's County Police Department, Maryland, spoke about requirements for first responders when dealing with digital evidence and what would be a "good seizure" of a computer involved in a crime. The first responder avoids letting the suspect or "bad guy" shut down the computer because evidence may be eliminated in that process. Wrong shutdown methods (or a bad floppy drive) may cause alteration of many files. The officer should then survey the scene for other media and preserve the integrity of the evidence until the time of

seizure. To obtain a baseline for the laboratory, the evidence computer's hard drive is "hashed" (i.e., a digital "fingerprint" of the hard drive is made to show that there has been no alteration since the time of seizure). The hashing makes a bit-stream image of the hard drive, revealing many sections of the drive where data and fragments are stored over which the user has no control. The police forensic unit then duplicates the evidence drive onto a clean second hard drive, allowing examiners to work from the copy for most procedures and touching the original as little as possible.

Occasionally, the evidence computer has been unavoidably altered between seizure and presentation at court. The law enforcement computer analyst then has to be able to testify in detail about which computer files were altered, using a knowledge of the Windows environment or other platforms.

Detecting Manipulation Artifacts in Digital Imagery

Russell H. Rosenthal, Special Agent and Certified Forensic Examiner, Computer Analysis Response Team, Washington Field Office, Federal Bureau of Investigation (FBI), and **Thomas Musheno,** Forensic Examiner, Forensic Audio, Video, and Image Analysis Unit, Investigative Technologies Division, FBI, discussed recent work with computer evidence. A computer may be a "victim" (e.g., robbed of data, hacked), an instrument of crime (e.g., used to prepare a ransom note), or evidence in some other association (e.g., used to store data that provide geographic or time evidence). Computer evidence should be regarded as "latent" evidence, which is easily altered and requires special precautions (both tools and training) for its preservation.

Special Agent Rosenthal said that nearly every kind of case, from bank fraud to terrorism, has some connection to digital

evidence now, due to the prevalence of computer communications and record-keeping. Until recently, only a few special investigators were qualified for computer analysis in forensic investigations, and little formal training was available for investigators who deal with computer evidence. The FBI's certified Computer Analysis Response Team (CART) agents receive training in the following:

- Personal computer repair and trouble-shooting.

- Net+ (Internet aspects).

- Data recovery and analysis (West Virginia Crime Center).

- Boot camp (FBI-specific procedures, with moot court presentations).

These agents are required to recertify annually and receive mentoring and on-the-job training in their specialties. They can use and read data from various media. To preserve computer drive evidence, CART agents will hash the drive by performing a math function that images the data stream. This process creates a "thumbprint" of the evidence drive that can be used in court to show that no tampering has occurred. FBI agents also use software called ILook, developed in the United Kingdom and purchased by the Internal Revenue Service (IRS), for data reduction and search. The program uses file extensions to organize files and folders.

Mr. Musheno presented information on detecting image manipulation. Images that show scars, moles, tattoos, or other characteristic marks are often used to identify individuals. Pictures also locate people or things geographically at a certain point in time, revealing significant connections. Skillful image manipulation is very difficult to detect. The examiner of digital evidence tries to find bad pixels or other metadata that indicate manipulation. "Artifacts"

(loss of pixel data caused by normal file compression in the communication process) create one of the biggest problems in distinguishing image manipulation from normal inherent anomalies. Camera identification offers another way to compare objects and locations. Irregular frame edges or dirty marks left from the roller are examples of particular camera patterns. The technician also watches for anomalies like wrong grain patterns in the film, incorrect shadowing, inconsistencies in size of relative objects, inconsistent degree of focus, or blurred edges.

Agents working with child virtual image pornography can search a database of images from known cases that was established during the 1970s and 1980s, when child pornography was legal in some parts of Europe. In some recent pornography cases, defense counsel claimed that child pornography figures were not real children but "virtual persons" and that there was no actual victim. Experts in computer graphics have said it is impossible to create "perfect virtual children," even with still images, much less moving ones. To create the animated figures for "Final Fantasy," a computer game, the file cost $150 million and consisted of 140,000 frames produced with custom-designed central processing units. Still, it was necessary for the filmmakers to use human models for the hands, face, and hair.

Stacey A. Levine, Computer and Telecommunications Coordinator, Organized Crime Strike Force, U.S. Attorney's Office, District of New Jersey, discussed the process of authenticating digital evidence during a trial. Prosecutors are seeing computer evidence in all kinds of cases, from child pornography to narcotics, kidnapping, and white-collar fraud. When agents are unable to find those who actually witnessed criminal activity, chatroom and e-mail files (although they are often anonymous) may authenticate a connection to a

suspect by means of circumstantial evidence.

One common defense is to seek exclusion of computer evidence as "hearsay." Generally, the court will admit an exception to this rule if the computer evidence is generated as a result of activities that are "relied on in the ordinary course of business," such as telephone toll records or bank account withdrawal records. Records that have been admitted under this exception to the hearsay rule include login records, dialup access time, addresses associated with Internet accounts, and file transfer records. Computer evidence must be authentic and must originate from reliable software (that is, the program that created the records has to be trustworthy and not prone to error).

Jury Understanding of Statistics: Plenary Panel

Moderator **Joe S. Cecil,** Project Director, Division of Research, Federal Judicial Center, Washington, D.C., noted that there has been a rich debate concerning limits on the understanding of juries, especially when expert witness evidence is very complex.

Jury Talk About Experts During Jury Deliberations

Shari Seidman Diamond, Howard J. Trienens Professor of Law and Professor of Psychology, Northwestern University Law School, spoke about an Arizona study she has been conducting with her colleagues Neil Vidmar and Mary Rose, in which juror deliberations were videotaped in 50 civil cases. The trials that were videotaped reflected the general makeup of most State civil court dockets: Half of the trials involved motor vehicle torts, a third involved other torts, and the remainder were medical malpractice and contract

cases. Awards varied from $0 to $2.8 million. More than 80 percent of the cases included at least one expert witness, and more than half involved opposing experts testifying on the same issue. Jurors were less interested in damage experts and more attentive to aspects of a decision that influenced a verdict or sentence.

Claims are often made that jurors are overawed by or dismissive of expert testimony. The jury deliberations in cases that were videotaped provided no support for such claims. Jurors discussed the experts frequently. Their discussion focused primarily on the content and plausibility of the expert testimony and the witnesses' apparent expertise and trustworthiness. Contrary to some expectations, jurors generally did not talk about the experts' appearance or clothing. They discussed the clarity of the presentations and evaluated the experts' credentials and relevant experience. In some cases, they commented on the amount that the expert was paid and offered negative comments about "hired guns." Jurors also responded negatively to what they perceived as unnecessary obfuscation (e.g., "Witnesses should speak English"), and they were critical of presentations that were too repetitious. Jurors in Arizona are permitted to submit questions to witnesses through the judge. The judge first determines whether the question is legally permissible. Jurors submitted questions to nearly half of the experts. Their questions revealed substantial sophistication. For example, jurors asked the meaning of "reasonable psychological probability" in connection with a diagnosis of posttraumatic stress disorder. In response to engineering testimony that estimated the speed of a moving vehicle at the time of an auto accident, a juror asked: "What is the error in your 10 mph estimate? Please give us a percentage or plus or minus number." The opportunity to submit questions, like the deliberations, provides evidence that the

jurors were engaged in serious efforts to understand and evaluate the expert testimony.

Statistical Evidence Requires Translation (or Exorcism?)

David A. Freedman, Professor of Statistics, University of California, Berkeley, discussed statistics in civil litigation and problems jurors have in understanding statistical evidence. He illustrated these problems using test results from undergraduate statistics courses at the University of California, where students typically have some mathematical preparation and receive 15 weeks of organized instruction. Jurors are unlikely to do better with the concepts than these students.

One important topic is study design, which generally determines the reliability of conclusions that may be drawn from the data. In an experiment, the investigator chooses which subjects receive treatment, for example; the remaining subjects are controls. Responses of the treatment group and the control group can be compared to determine the effect of treatment. Often social circumstances do not allow experiments to be done, and observational studies must be used instead. In an observational study, the subjects themselves select the treatment or control conditions.

The chief problem with observational studies is confounding (i.e., the treatment and control groups differ in some important way besides treatment). In that case, association does not necessarily imply causation. For example, an association exists between smoking and cirrhosis. Smokers have a higher rate of cirrhosis than nonsmokers, but smoking is not the cause. Alcohol consumption is the confounding variable: Smokers drink more on average than nonsmokers, and alcohol damages the liver. Statisticians recommend assigning subjects to treatment or control groups using a well-defined random mechanism rather than the judgment of the investigators, and this too is a difficult idea for students.

Only about half the students understood the bias that would be caused by failure of randomization in a Canadian study of mammography (screening for breast cancer by x-rays). The idea is that screening leads to earlier detection and more effective treatment. In this study, death rates from breast cancer were compared in the treatment group—invitation to screening—and the control group. Apparently, nurses assigned high-risk women to the treatment group. This would bias the study against screening, but students think the bias goes the other way: Putting high-risk women into the treatment group allows screening to demonstrate its merits. The problem is that doctors cannot determine which women would have lived had they been screened—that is why experiments are needed in the first place.

Another great mystery is hypothesis testing. Significance levels measure the likelihood of the data given the null hypothesis, not the likelihood of the null hypothesis given the data. Mixing up these two likelihoods is the "transposition fallacy," called the "prosecutor's fallacy" in the context of DNA identification. At best, the odds ratio answers the question, "If the defendant is innocent, what are the odds against getting a match?" The odds ratios cannot answer the real question, "What are the odds against the defendant being innocent?"

Statistics Jurors Cannot Do Without

Lawrence M. Solan, Director, Center for the Study of Law, Language, and Cognition, Brooklyn Law School, spoke about poorly understood probabilistic information in courts. Not only judges but also people in general misunderstand base rates.

People tend to make decisions according to their own experience, regardless of research. For example, in a case involving handwriting, the author of a threat and the source of questioned handwriting made similar use of spaces and similar punctuation mistakes and spelling errors. However, no information was presented to the court about the general frequency of these characteristics in the New Jersey population. No baseline showed, for example, how many people usually make the same kind of punctuation errors. Jurors typically do not understand how to discount evidence based on a statistical base rate. Sometimes, a judge will allow an expert to point out similarities and differences without expressing opinions about the evidence. The legal system must struggle with the concept because everyday reasoning does not "ask the baseline questions." Psychiatric evidence is another example where baseline statistical information is critical both for adversaries in the court and for the jury.

Bioengineering and probabilistic analyses are difficult material even for judges, who at least have recourse to the opposing attorneys to clarify issues. Since the 1970s, Dr. Solan noted, jury trials have declined in number, an effect probably related at least in part to the educational challenge posed by modern science issues in the courtroom.

Call-for-Papers Presentations

The Role of Meta-Analysis in the Legal System

Jeremy A. Blumenthal, Psychology Department, Harvard University, discussed the common social science methodology of meta-analysis, which pools and statistically analyzes the results of all existing empirical studies in a research area. More than a simple recalculation of multiple studies' data, meta-analysis can provide courts and legislators with the most current and thorough summary of the research through quantitative rather than narrative synthesis. Researchers or expert witnesses can weight each study for methodological quality, internal validity, sample size, and other factors and give a better indication of the state of research. By using the whole body of available research on a topic, meta-analysis increases statistical power and addresses methodological flaws in individual studies. Courts and policymakers receive more background for a decision than can be provided by discussions of single studies or citations to law review articles.

By collecting current information, balancing presentations with opposing views, and addressing any limitations on the existing research, meta-analysis discovers the general state of knowledge in a research area. It highlights practical implications for courts and legislators, and it can empirically address assumptions made in the legal system. The analysis uncovers moderator variables and shows when empirical evidence "can be more than the sum of its parts."

Voodoo Science by Default

Peter R. De Forest, Professor of Criminalistics, John Jay College of Criminal Justice, New York, discussed crime scenes as examples of "scientific problems" and said scientific expertise works to reconstruct, interpret, and (perhaps most important) integrate information at a crime scene. Forensic science programs have, in many areas, developed without formal standards, relying on laboratories' quality assurance and on-the-job training. He noted that not only scientists use scientific methods; however, some organizations in the last decade have trained people to be "crime scene experts" without enough qualification to permit interpretation of evidence. An example could be some people trained through the

many State-run fire academies. To understand fire scene evidence properly, a person needs a thorough background in fire dynamics and chemistry. Unfortunately, sometimes people who lack this background are hired as "experts" simply because they are persuasive speakers in the courtroom.

Physical evidence at a crime scene records human interactions and different relations of energy and matter that "encode" information about the crime. Asking the right questions is a key element. Many times, people who do a routine job at a crime scene, just following protocols, lack the scientific knowledge needed to make specific hypotheses and frame the right questions. The process used to get the facts and information is the important thing, rather than certain accoutrements of a scientist. In court, expectations of certainty are often unrealistic. Science gives a "best answer" rather than a certain one, sometimes suggesting several interpretations that can be made of the available information.

Inconsistency in Eyewitness Testimony: What Does It Really Tell Us?

Ronald P. Fisher, Professor of Psychology, Florida International University, discussed his research on the accuracy of eyewitness testimony and the claim that eyewitness testimony should be discredited because of inconsistent witness memory of event components. This strategy assumes that correct memory of particular components of the event will predict accurate identification (of a suspect, for example). Attorneys typically believe that witnesses who give inaccurate information cannot be relied on to make correct identifications. A crime event, however, is complex and recorded in a "holistic" way.

In Dr. Fisher's study, a simulated crime event was witnessed, and witnesses were

given tests to gather their knowledge of specific characteristics of the event. In the first set of results (as expected), statements given more consistently did predict more accuracy. The study then asked whether inconsistency in three or four statements would predict global inaccuracy. Here, the correlation was weak. Inconsistent witnesses were not much less accurate than consistent witnesses. Some other dimensions, such as the witness's motivation to lie, may be essential considerations or more predictive of inaccuracy. Witnesses were tested twice and up to 2 weeks after the simulated event. Conventional investigations, in contrast, often continue for several months or even more than a year after a crime.

Judges who instruct jurors to use consistency as a major factor to try to assess whether or not a witness is accurate are, in some ways, encouraging a juror to decide based on something invalid, at least in terms of research tests. Inconsistency does not seem to be a very good predictor of accuracy in identification.

The research also showed that some factors have opposite effects on consistency and accuracy. For example, when time between the event and the test was varied, a longer period led to decreased accuracy but increased consistency. The memory "solidifies" but may be less accurate. This makes it clear that two different concepts are measured. Also, encouraging guesses clearly increases inaccuracy. But inconsistency in the witness's memory alone should not discredit all testimony.

Jurors' Comprehension of Contested DNA Evidence: A Case Study

William C. Thompson, Professor, Department of Criminology, Law, and Society, University of California, Irvine, described a southern California murder trial that turned on conflicting interpretations of DNA

evidence. Dr. Thompson had been cocounsel for one of two brothers accused of killing a convenience store clerk during a robbery. Two expert witnesses interpreted the DNA evidence in opposing ways, and the jury was asked to decide the case based on their testimony. Posttrial interviews revealed how the jurors had understood (and misunderstood) the disputed scientific evidence.

A surveillance videotape showed the clerk grapple with one of the robbers before the other robber shot him. The shooter wore a white cap, which fell off when he departed. Tests revealed foreign DNA under the victim's fingernails that matched neither defendant in the case. The defense argued this foreign DNA was from one of the robbers; the prosecution contended it was from an unknown person unrelated to the crime. DNA tests also found a mixture of DNA (profiles from more than one person) on the white cap. One defendant was "excluded" as a possible contributor to this mixture. The other defendant was also excluded *if there were two contributors to the DNA on the hat* but was included as a possible contributor *if there were three or more contributors to the DNA on the hat*. The prosecution contended that there were three contributors to the DNA on the hat, and hence that the DNA evidence incriminated the defendant; the defense contended there were two contributors and hence that the DNA exonerated him.

At issue was the interpretation of electropherograms produced by the ABI 310 Genetic Analyzer (currently used by about 85 percent of crime laboratories). The prosecution expert argued that peak-height disparities in the electropherograms indicated the presence of three contributors. The defense expert contended that the peak height disparities were consistent with 2 contributors and argued that the theory of 3 contributors was implausible given the small number of alle-

les (3 or 4) detected at each of 13 loci. The defense expert presented likelihood ratios purporting to show that the observed results were 250 times more likely under the theory of 2 contributors than 3 contributors. The defense expert also challenged the government's random-match probabilities. The prosecution sought to discredit the defense expert because his experience primarily involved animal rather than human DNA. The defense attacked the prosecution expert on grounds that he lacked a 4-year college degree.

The jurors deliberated for 6 days before deadlocking 9 to 3 in favor of conviction. In posttrial interviews, jurors expressed anger at being asked to resolve such a difficult scientific issue. They were suspicious of the defense arguments because they had heard that DNA evidence is virtually infallible and definitive. They dismissed some defense arguments based on faulty assumptions. For example, they assumed that the foreign DNA under the victim's fingernails was from hot dogs he was preparing before the crime (even though the experts had testified that the tests are human specific). And they mistakenly assumed that peak-heights correspond to the number of contributors, rather than the quantity of DNA (an assumption that supported the prosecution claim of three or more contributors). They did not remember or understand the defense argument about likelihood ratios.

Most DNA cases are more clearcut, Dr. Thompson said, but when small and mixed samples are involved, there are often difficult interpretive issues that could be considered too complex for a lay jury. Jurors give so much credibility to DNA evidence that it is difficult for defense counsel to challenge it in cases where it is less than definitive. These findings argue for stricter admissibility standards to ensure that the DNA evidence that goes to the jury warrants the confidence jurors are likely to have in it. When such issues are

litigated, Dr. Thompson recommended introducing more experts (a single one on each side seemed to encourage doubts in the jurors) and taking more time to educate jurors to the nature of technical problems that can occur. He also noted the importance of seating at least some jurors who have scientific backgrounds. Scientific evidence can contribute greatly to the accuracy of criminal trials but also has the potential to be misleading if not handled carefully.

Saturday, October 5

Fingerprints—Making Sense of Forensic Science: Plenary and Roundtable

Moderator **David Faigman,** Professor of Law, Hastings College of Law, University of California, San Francisco, said current treatment of fingerprint evidence in courts may form a keystone for other kinds of forensic science evidence, such as arson, firearms/toolmarks, trace evidence, handwriting, and more.

Forensic Fingerprints

Sandy Zabell, Professor, Department of Mathematics and Statistics, Northwestern University, spoke about the historical background of fingerprint use for identification of individuals. The Chinese have used fingerprinting for hundreds of years to verify documents, but the use of fingerprinting for personal identification arose in the context of mobile urban societies. A French police officer, Bertillon, developed a system of taking 11 different anthropomorphic measurements and photographs in 1883 for the purpose of identifying individuals/suspects. Easily accessible records were kept on cards, but "uniqueness" was not considered an issue. A disadvantage to the system was that examiners needed careful training and oversight to take measurements correctly. Also, the system was calibrated for Parisians, which made it difficult to apply in other settings (for example, by the British operating in India during that time period).

Other well-known identification systems were developed by William Herschel (1858); Henry Faulds (1889); Francis Galton (1889), who was famous for the first fingerprint classification system that used arches, loops, and whorls as marks of identification; and Edward Henry (1895), who first matched "clean sets" and whose classification system quickly became the standard. Recent questions about partial prints and the introduction of applied probability to the examination of fingerprint evidence have overshadowed the use of fingerprint identification for suspects in cases at law.

Forensic Individualization of Images

John R. Vanderkolk, Laboratory Manager, Fort Wayne Regional Laboratory, Indiana State Police, described the agency's formal method of critical inspection, which is known as ACE+V:

- **A.** Analysis determines essential details of the fingerprint: the surfaces involved, processing techniques used, and quality and quantity of detail in the images.

- **C.** Comparison is made noting similarities or differences between the two images, recognizing that there are no perfect images; variations in appearance and distortions are noted.

- **E.** Evaluation is the determination of the significance of the analysis and comparison. The standard used to determine agreement of detail is based on the

quality and quantity of the detail in the two images.

- **V.** Verification is made: A second examiner repeats the process of ACE and determines whether he or she can replicate ACE and reach the same conclusion as the first examiner.

Different levels of features can be observed on all friction ridge skin. Each latent print (image) represents a partial record of friction ridge skin. All images are partial. It is necessary to reach agreement of unique details in the images to determine that they came from the same source. At level one, the image is recognized as a fingerprint based on direction of ridge flow; at level two, it shows actual ridge path, length, division or bifurcation, and width; and at level three, it shows contour and texture of ridges and pore position. All levels of detail present in the images can be measured comparatively. The clearer the detail, the more power or significance that detail has. Mr. Vanderkolk emphasized the examiner's need to understand the uniqueness and durability of the actual friction ridge skin or source of fingerprints. After that, the examiner needs to understand the relationship between quality and quantity of the levels of detail in the two images. As quality of the images increases, the requirement for quantity decreases. As quantity of the images increases, the requirement for quality decreases.

Testing: Is It Reliable?

William J. Tilstone, Executive Director, National Forensic Science Technology Center, Largo, Florida, said the main question to consider with reference to fingerprints and other trace evidence is whether the evidence is reliable. *Daubert* established a role for the judge in determining that material taken into evidence is both relevant and reliable.

It is hard to explain what is scientific: Kuhn's behavioral model describes the scientific method as a "discrete group [of qualified persons] following some rules and conducting experiments." People make the mistake of confusing "scientific" with reliable. Academic degrees and publications are peripheral considerations; they do not always indicate expertise in scientific method and are sometimes featured too prominently.

In the scientific method, the examiner seeks to find out how nature works, using a "test, refine, test again" cycle. Reliability is not the defining characteristic of science: Airplane pilots and bankers are reliable, but not necessarily scientists. The scientific community has definite rules about reliability of *testing*; these are well defined and articulated and should be addressed in debates about fingerprints. Laws of nature have emerged that withstand the test of time (for the court, this compares to a *Frye* rule test).

Error rates become important because science is not exact. When "objective tests" are discussed, this means they are conducted with validly designed controls and documented (i.e., appropriately trained staff will obtain the same results within defined limits). Validation may also be made by comparison to certified reference materials (where available) and published standards from the scientific community. Debate on scientific evidence focuses more productively on ways to show adherence to standards, practitioner reliability testing, and development of certified reference materials.

Appropriate Validation for Nonscientific Fingerprint Expertise

Edward J. Imwinkelried, Professor of Law, University of California, Davis, said forensic fingerprint evidence should be admitted as technical expertise, not as a

"science." The *Daubert* standard recognizes the limitations of scientific enterprises. It would therefore be anomalous to make scientific testing the litmus test for the admissibility of all types of expert testimony. "If we prescribe a general 'evidence rule' requiring the proponent whenever possible to present scientific evidence," said Professor Imwinkelried, "we handicap the court." As regards questioned documents, for example, experienced document examiners have accurately identified sources with an error rate 13 times lower than that among laypeople. Barring questioned-document testimony will force the courts to rely on lay testimony, which is demonstrably less reliable.

Although it is clear that forensic trace evidence should not be excluded from the court, it is much harder to figure out what nonscientific, technical standards should be used for this evidence. Professor Imwinkelried considered the example of an undercover agent's knowledge of code words used in the illegal drug trade. Federal Rule of Evidence 702, he said, should not be construed in a narrow scientific sense but, rather, in a broader rationalist sense. Showing that a process produces an accurate result qualifies it as admissible, whether or not it is "science."

Even if the courts admit forensic science testimony, the opposing attorney can effectively attack the weight of the testimony by demonstrating the lack of supporting scientific research. The availability of that weight attack to the opponent helps keep forensic scientists honest and gives them an incentive to conduct better research. Currently, many evidence fields are "scrambling to collect data," but some observers in the audience questioned whether, if no court requirement emerged, the field would go back to the pre-1993 status quo. Imwinkelried added that to ensure that the jurors do not attach undue weight to nonscientific testimony, the judge can give the jury a cautionary instruction, advising them that the basis of the expert's testimony is simply experience, not full-fledged science. Accuracy of the test or method's results should be the most important consideration in determining the admissibility of evidence in court. The accuracy of FBI tests in *Plaza*[8] was staggering. In the International Association of Identification certification tests, 58 percent of examiners had not made a single error. Their techniques accurately led to the inferences they claim they can make. The foundational showing was not a conventional scientific demonstration, but case law, statute, and policy do not treat conventional science as the solitary criterion. The social "faith" in science is based on a huge body of results in society produced by scientific method. Professor Imwinkelried said, "In the microcosm of the courtroom as well, the focus should also be on results."

Why Testing, Good Testing, and the Results of Testing Are the Touchstones of Good Science and Good Judicial Gatekeeping

Michael J. Saks, Professor of Law, Arizona State University, said *Daubert* asks judges to refrain from admitting evidence unless it is based on valid methods (i.e., the word of an expert alone is not sufficient). A judge should be able to see and evaluate serious research that empirically tests claims about expert fingerprint evidence. Challenges to fingerprint evidence have now been brought and judicial opinions have been written in about a dozen cases. Federal judges have concluded in all cases that the expert evidence was admissible. But if we read what the judges said in those opinions, we discover that not one of those courts actually found that the proponents of the expert testimony could satisfy the *Daubert* test. Instead, the opinions found various ways to evade *Daubert* and make sure the evidence was

admitted. For example, some courts just refused to hold a *Daubert* hearing, others reversed the burden of proof, and some just concluded they would admit the evidence without any explanation. All the courts so far have ignored the "task at hand" requirement. Some courts have relied on general acceptance, and others have applied the *Kumho Tire* decision as if it narrowed the ground for scrutinizing evidence.

Professor Saks said he regrets that courts could not limit experts' statements in more precise ways, rather than merely admitting or excluding them. Courts should be better able to welcome good science but to filter out statements that go beyond what has come to be known to be true through sound research. As more background research is conducted, statements in court will become tempered and accurate, rather than "merely believed." That research is more likely to be done, however, if judges do the policing of expert evidence that is required of them by *Daubert* and *Kumho Tire*.

Open Discussion

Courts are not designed to thrash out scientific issues; the adversarial procedure prevents this. Too often, private industry, with a vested interest in selling particular equipment or methods, becomes involved in the testing issue. Professor Imwinkelreid said, however, that a lack of research could threaten the forensic science community's credibility. The group discussed experts and evidence testing in connection with defendants who are indigent (85 percent). How can empirical studies address a lack of data in such cases? Judges usually cannot order general studies on better methodologies in the required time frames.

Audience members brought up the lack of research budgets for local, public

forensic science laboratories. Legislative representatives and public agencies need to understand that academic institutions will not conduct research without support. Similar questions have been raised concerning handwriting, footwear, firearms, and other pattern recognition evidence. Courts cannot "throw out the only available evidence" just because empirical studies of methods are not available.

Funding for methods research competes with other administration priorities. In the United Kingdom, the Home Office destaffed an existing forensic research laboratory that published 30 studies a year to support the growing DNA database. In recent reviews of six States' crime laboratories, research found the quality of service to be closely related to the level of government executive to whom the laboratory director reported. If the laboratory director reported to the State attorney general, the work tended to be well researched and documented; if, however, the laboratory director reported to a subordinate law enforcement official, problems were seen more often. Time pressure is a major factor as well; most laboratories would prefer to see money used to enable them to obtain speedier results for the justice process.

Ms. Ballou mentioned that the National Institute of Standards and Technology offers publications on methodologies and images as well as a 12- to 15-year history of work on fingerprint evidence. Law enforcement would benefit from an effort to compile this reference material. Professor Saks said existing studies do not sufficiently address probability (i.e., the risk of coincidental matches). Assumptions about rarity or uniqueness are important for the scientific rigor of the comparison process. Use of careful (blind) protocols, as has been done recently for eyewitness evidence, is a specific step toward a more objective process. In casework, examiners need to

receive only the information about the case that is necessary to perform their examination, with no extraneous information. At the Home Office in the United Kingdom, applications of Baysean or random-match software are used to evaluate some trace evidence decisions. Similar research methods can apply to signatures. The full meaning of "good testing procedures," however, remains an open question.

Keynote Address: Knowledge, Power, and the Evolving Role of Scientific Evidence

Judge Gerald T. Wetherington, Vice President, Duke University Private Adjudication Center, and Partner, Wetherington, Klein & Hubbart, P.A., spoke about the mental struggles by juries, judges, and expert witnesses as they strive to reach the right decision in a complex case. He described an example of a murder case in which a coworker of the victim testified to hearing the defendant threaten to kill her when he called her at her place of employment. It was the second time he had called that day, and her coworkers observed that she was very upset from the incident of his first call.

Florida has strict laws against listening to telephone conversations without consent of the parties, subject to particular exceptions. One exception concerns calls to business phone numbers related to matters of concern to the business. Judge Wetherington, after much thought, decided to admit the crucial evidence since the first call to the victim-employee had a large impact on her performance; the other employee, observing this, could then consider the second call to have an impact on the business. The State's Supreme Court affirmed his decision, but

he felt that, had he agreed to exclude the evidence, the matter would never have been reviewed because the defendant could not be placed in double jeopardy.

Judge Wetherington quoted an inscription shown in the Miami-Dade courts: "We who labor here seek only the truth." He noted that, as long as human beings (whether judges, jurors, or scientific experts) are involved in reaching a decision, there is bound to be a need to set aside subjective factors that may exercise an influence over that decision: "How will the decision be received professionally? Will it make me look good? Will it bring me more or better work? Or, will someone hate me for this decision?" Forensic and technical experts, like judges, take an oath to tell the truth, and, at a societal level, they play increasingly critical roles in the effort to influence the conduct of masses of people and to preserve social order while protecting individual rights. This is, he said, one of society's most important functions.

Valuation for objectivity in the search for truth, devised by the ancient Greeks who developed science, makes the "light" of expert witnesses seem determinative in a complex case and gives them great power. They are credited with perhaps "mysterious" knowledge beyond the common knowledge of the factfinders, so they must confront all factors that play a role in their decision, including observer bias. They, like the judge, must undergo internal scrutiny of their motives to reach a standard that is (quoting Justice Cardozo) "something stricter than the morals of the marketplace." The expert has to ask himself or herself, "What is my duty? What or whom do I serve? And, am I a trustee for objective truth?" The only dependable protection is a commitment to principles of integrity regardless of adversity.

Case Scenario Involving Hair Evidence: Microscopic Examination—How Much Can Hair Tell?

Judge André Davis, U.S. District Court, District of Maryland, moderated this scripted legal scenario on hair evidence. **Rockne P. Harmon,** Senior Deputy District Attorney, Alameda County, Oakland, California, played the role of prosecutor. **Max M. Houck,** Director, Forensic Science Initiative, West Virginia University, played the first expert witness. **Barry C. Scheck,** Professor, Benjamin N. Cardozo School of Law, New York, played the defense counsel. **Terry Melton,** President and Chief Executive Officer, Mitotyping Technologies, Pennsylvania, played the role of second expert witness.

In the hypothetical case, the victim was found murdered in her own home, naked, on a carpeted floor. She was strangled; there were no indications of sexual assault. Neighbors heard screams at about 2:30 a.m. A registered sex offender, J. Franklin, lived near the victim. Witnesses who were shown a photo lineup sequentially identified J. Franklin as the person they saw leaving the victim's property near the time of the murder. Franklin denies knowing the victim or ever having been in her apartment. He has filed an alibi and has three witnesses. Franklin's previous conviction was for sexual assault with choking. The judge has requested that the first expert witness give a neutral tutorial to the jury concerning hair evidence in the case.

Defense counsel objected, for the record, to the lack of notes of test results given to the defense, to the lack of proficiency testing for the laboratory scientist, to the refusal to allow DNA testing, and to the lack of judicial opinion concerning rejection of the defense's challenge to admissibility

of hair evidence under the *Daubert* and *Frye* case standards. The expert witness took the stand and gave his oath.

First Expert Witness: Trial on the Merits—Tutorial to Jury

The expert witness, Mr. Houck, gave an overview of the hair examination processes. Anthropology, comparative biology, microbiology, and histology (human tissue study) all serve as bases for forensic hair evidence. Hair is significant because of the Locard exchange principle, which states that every contact leaves a trace. Direct or indirect transfers of hair indicate and connect persons and things in particular locations (crime scenes, for example). Transfer of hair from one person to another would be a direct transfer; transfer of hair to a car seat and then to another person would be indirect. Mr. Houck showed slides of the morphology of hair: cuticle, medulla, and cortex (the location of elongated cells containing pigment). Animal hair is very different from human hair through variation in pigmentation and banding, for example. The cuticles of animal hairs vary by species. Estimations of human ancestry (African, Asian, or European) from microscopic hair examinations are possible but are only estimations.

Microscopic comparison is typically carried out only on human head or pubic hair. Examiners use a comparison microscope (two optically joined instruments with a split-screen view used to compare two samples side by side). To give a good representation, the sample must contain at least 25 to 50 hairs, both combed and plucked. Actively growing hairs look quite different from those that have been naturally shed. The comparison looks at the root, shaft, and tip in terms of all characteristics (diameters, structure, color). A point-by-point detailed comparison is made between the unknown and known samples. Of the samples found at the

crime scene, in the instant case, six samples could have been from the defendant and the four hairs found on the victim's back could not have been from the defendant (but were all from the same person).

Defense counsel then questioned the witness about alternative hair extraction testing using mitochondrial DNA (mt-DNA). In 9 of 80 FBI cases where there had been an association based on hair, mt-DNA testing resulted in an exclusion of the defendant. The witness agreed that, if DNA testing were done and Mr. Franklin was excluded, he could not be the source of the sample hairs.

Examiners' experience varies greatly: Some have seen thousands of samples and some have seen many fewer. State laboratories may have less time and expertise than larger commercial laboratories. The significance of an association between hair samples falls short of positive identification. When examiners disputed hair findings and an outside laboratory was commissioned to review the testing, defense counsel noted, in 50 percent of cases new findings and (ultimately) a conclusion to exclude the defendant resulted.

Open Discussion [Out of Role]

Hair evidence is not easily quantifiable or digitized. Length, thickness, and pigment patterns are hard to capture and standardize. As a result, proficiency testing for forensic hair analysis cannot be mass produced. Mr. Scheck noted that, in mt-DNA testing, reliability and validity can be more readily achieved. Tutoring a jury in the meaning of applicable analyses and tests may bring more understanding. Mr. Harmon noted that a prosecutor has to build on what is available; there may be obstacles to obtaining mt-DNA tests.

Second Expert Witness: Postconviction Motion for DNA Analysis

The second expert, Dr. Melton, described her background. She has published papers on mt-DNA population genetics, applied forensic uses of mt-DNA, and population histories in Southeast Asia, Taiwan, and Kenya. The human mt-DNA sequence has been known for about 19 years. Mt-DNA is an abundant molecule found in hair, bone, and other tissues (nuclear DNA is less abundant). The Armed Forces uses it to identify war dead and MIAs from Korea and Vietnam, and it has been used in anthropology to study origins and migrations of human populations. In medical genetics, the mitochondrion is also very important as the organelle that provides energy to the cell. A change in its mitochondrial DNA genes may result in a severe mitochondrial disorder.

Defense counsel asked about validation studies for a DNA laboratory. Each laboratory must carry out validation studies for itself and look for limitations in its methods. Federal DNA advisory boards have established standards for laboratories handling DNA. Dr. Melton's laboratory has performed many hair extractions (this is 85 percent of its business). Since 1999, she has participated in more than 300 cases involving about 1,200 samples. In 12 admissibility hearings involving Dr. Melton, the courts decided to admit the mt-DNA evidence.

Dr. Melton gave a brief tutorial on the mt-DNA molecule and its analysis, including PCR amplification and DNA sequencing. She demonstrated DNA sequences that showed an exclusion, an inconclusive comparison, and a heteroplasmic position that would still result in a failure to exclude. Maternally related individuals will have the same mt-DNA sequence. Heteroplasmy, defined as the presence of more

than one type of mt-DNA in an individual, occasionally occurs and would still be interpreted as a failure to exclude in most cases. Use of microscopic comparisons on hair can narrow the field of samples to be submitted for DNA testing. Microscopy can make compelling evidence in combination with other things, but an investigator should not assume knowledge of a match based on microscopy, and mt-DNA testing of hairs is essential.

Dr. Melton discussed the statistical significance of mitochondrial DNA matches. The FBI's forensic database (about 4,800 mt-DNA sequences) is used for North Americans. An examiner can search the database for a profile to estimate the frequency of a mt-DNA type, see how common or rare it is, and place a statistical value on it. If the type were not in the database at all, then 99.94 percent of North Americans would not be expected to have this mt-DNA type, with 95 percent confidence.

Open Discussion [Out of Role]

The group spoke about postconviction DNA cases. Mr. Scheck noted that 27 States currently have statutes supporting postconviction testing. He described the difficulty of searching State transcripts to prepare an application for postconviction DNA testing. It may take 4 years to prepare evidence for such a case. In 75 percent of the cases, the evidence has been lost or destroyed, but where the evidence can be found, about half the time it turns out to be favorable. Currently, 112 postconviction DNA tests have exonerated people in prisons.

Closing Remarks

Lisa Forman, Director, Investigative and Forensic Sciences Division, Office of Science and Technology, National Institute of Justice, U.S. Department of Justice, noted that science and law share the goal of reaching sound conclusions based on what can be investigated effectively. Technical investigative methods are good complements to scientific procedure, but discussion too often focuses on contentions between disciplines. Fingerprints, digital images, eyewitness identification, firearms traces, and other evidence play critical roles in crime investigation and must receive research attention. Dr. Forman invited contact and asked participants to inform NIJ about research they viewed as necessary under the *Daubert* guidelines. She thanked all participants on behalf of Director Hart.

Notes

1. *Brady* v. *Maryland*, 373 U.S. 83 (1963).

2. *State* v. *Zain*, 207 W. Va. 54, 528 S.E. 2d 748 (1999).

3. "Preliminary Gilchrist Reports," Oklahoma State Bureau of Investigation, on work of police chemist Joyce Gilchrist.

4. *Daubert* v. *Merrill Dow Pharmaceuticals, Inc.*, 509 U.S. 579 (1993).

5. *Frye* v. *United States*, 293 F. 1013 (D.C. Cir. 1923).

6. *Westerfield* v. *Superior Court of San Diego County*, 119 Cal. Retrial 588 (2002).

7. *Kumho Tire Co.* v. *Carmichael*, 526 U.S. 137 (1999).

8. *United States* v. *Plaza*, 179 F. Supp. 2d 492 (E.D. Pa), *vacated*, 188 F. Supp. 2d 549 (E.D. Pa. 2002).

About the National Institute of Justice

NIJ is the research, development, and evaluation agency of the U.S. Department of Justice. The Institute provides objective, independent, evidence-based knowledge and tools to enhance the administration of justice and public safety. NIJ's principal authorities are derived from the Omnibus Crime Control and Safe Streets Act of 1968, as amended (see 42 U.S.C. §§ 3721–3723).

The NIJ Director is appointed by the President and confirmed by the Senate. The Director establishes the Institute's objectives, guided by the priorities of the Office of Justice Programs, the U.S. Department of Justice, and the needs of the field. The Institute actively solicits the views of criminal justice and other professionals and researchers to inform its search for the knowledge and tools to guide policy and practice.

Strategic Goals

NIJ has seven strategic goals grouped into three categories:

Creating relevant knowledge and tools

1. Partner with State and local practitioners and policymakers to identify social science research and technology needs.
2. Create scientific, relevant, and reliable knowledge—with a particular emphasis on terrorism, violent crime, drugs and crime, cost-effectiveness, and community-based efforts—to enhance the administration of justice and public safety.
3. Develop affordable and effective tools and technologies to enhance the administration of justice and public safety.

Dissemination

4. Disseminate relevant knowledge and information to practitioners and policymakers in an understandable, timely, and concise manner.
5. Act as an honest broker to identify the information, tools, and technologies that respond to the needs of stakeholders.

Agency management

6. Practice fairness and openness in the research and development process.
7. Ensure professionalism, excellence, accountability, cost-effectiveness, and integrity in the management and conduct of NIJ activities and programs.

Program Areas

In addressing these strategic challenges, the Institute is involved in the following program areas: crime control and prevention, including policing; drugs and crime; justice systems and offender behavior, including corrections; violence and victimization; communications and information technologies; critical incident response; investigative and forensic sciences, including DNA; less-than-lethal technologies; officer protection; education and training technologies; testing and standards; technology assistance to law enforcement and corrections agencies; field testing of promising programs; and international crime control.

In addition to sponsoring research and development and technology assistance, NIJ evaluates programs, policies, and technologies. NIJ communicates its research and evaluation findings through conferences and print and electronic media.

To find out more about the National Institute of Justice, please visit:

http://www.ojp.usdoj.gov/nij

or contact:

National Criminal Justice
  Reference Service
P.O. Box 6000
Rockville, MD 20849–6000
800–851–3420
e-mail: *askncjrs@ncjrs.org*

www.ingramcontent.com/pod-product-compliance
Lightning Source LLC
Chambersburg PA
CBHW081841170526
45167CB00007B/2867